Roskosch Andrea

The influence of macrozoobenthos in lake sediments

Roskosch Andrea

The influence of macrozoobenthos in lake sediments

Influence on hydrodynamic transport processes and biogeochemical impacts

Südwestdeutscher Verlag für Hochschulschriften

Impressum/Imprint (nur für Deutschland/only for Germany)
Bibliografische Information der Deutschen Nationalbibliothek: Die Deutsche Nationalbibliothek verzeichnet diese Publikation in der Deutschen Nationalbibliografie; detaillierte bibliografische Daten sind im Internet über http://dnb.d-nb.de abrufbar.

Alle in diesem Buch genannten Marken und Produktnamen unterliegen warenzeichen-, marken- oder patentrechtlichem Schutz bzw. sind Warenzeichen oder eingetragene Warenzeichen der jeweiligen Inhaber. Die Wiedergabe von Marken, Produktnamen, Gebrauchsnamen, Handelsnamen, Warenbezeichnungen u.s.w. in diesem Werk berechtigt auch ohne besondere Kennzeichnung nicht zu der Annahme, dass solche Namen im Sinne der Warenzeichen- und Markenschutzgesetzgebung als frei zu betrachten wären und daher von jedermann benutzt werden dürften.

Coverbild: www.ingimage.com

Verlag: Südwestdeutscher Verlag für Hochschulschriften GmbH & Co. KG
Heinrich-Böcking-Str. 6-8, 66121 Saarbrücken, Deutschland
Telefon +49 681 37 20 271-1, Telefax +49 681 37 20 271-0
Email: info@svh-verlag.de

Approved by: Berlin, HU, Diss., 2012

Herstellung in Deutschland:
Schaltungsdienst Lange o.H.G., Berlin
Books on Demand GmbH, Norderstedt
Reha GmbH, Saarbrücken
Amazon Distribution GmbH, Leipzig
ISBN: 978-3-8381-3099-6

Imprint (only for USA, GB)
Bibliographic information published by the Deutsche Nationalbibliothek: The Deutsche Nationalbibliothek lists this publication in the Deutsche Nationalbibliografie; detailed bibliographic data are available in the Internet at http://dnb.d-nb.de.

Any brand names and product names mentioned in this book are subject to trademark, brand or patent protection and are trademarks or registered trademarks of their respective holders. The use of brand names, product names, common names, trade names, product descriptions etc. even without a particular marking in this works is in no way to be construed to mean that such names may be regarded as unrestricted in respect of trademark and brand protection legislation and could thus be used by anyone.

Cover image: www.ingimage.com

Publisher: Südwestdeutscher Verlag für Hochschulschriften GmbH & Co. KG
Heinrich-Böcking-Str. 6-8, 66121 Saarbrücken, Germany
Phone +49 681 37 20 271-1, Fax +49 681 37 20 271-0
Email: info@svh-verlag.de

Printed in the U.S.A.
Printed in the U.K. by (see last page)
ISBN: 978-3-8381-3099-6

Copyright © 2012 by the author and Südwestdeutscher Verlag für Hochschulschriften GmbH & Co. KG and licensors
All rights reserved. Saarbrücken 2012

Zusammenfassung

Ziel dieser Dissertation ist, Wissensdefizite im Bereich der Bioirrigation von Süßwassersedimenten abzubauen. Als Untersuchungsorganismus wurde *Chironomus plumosus* ausgewählt weil diese weit verbreitete Larve ihre U-förmigen Röhren mit Überstandswasser durchspült und so Nahrung aus dem Wasser filtriert.

Um Bioirrigation in Wohnröhren mit einem Durchmesser von ca. 1,7 mm zu untersuchen, wurden geeignete Messmethoden entwickelt, mit denen für das 4. Larvenstadium die Parameter Fließgeschwindigkeit (14,9 mm/s), Pumpzeit (33 min/h) und Pumprate (61 ml/h) gemessen wurden. Bei einer Populationsdichte von 745 Larven/m^2 kann somit ein Wasservolumen äquivalent zum Volumen des Müggelsee in Berlin, innerhalb von 5 Tagen durch das Sediment gepumpt werden.

Die Positronen-Emissions-Tomographie wurde für die Sedimentanalytik adaptiert und der Transport im Porenwasser analysiert. Mit den Untersuchungen wure gezeigt, dass auch in schlammigen Seesedimenten ein advektiver Transport durch Bioirrigation verursacht wird, der nicht zu vernachlässigen ist.

Steigende Temperaturen resultieren aufgrund steigender Fließgeschwindigkeit in einen signifikanten Anstieg der Pumprate sowie der Eintragsrate von Überstandswasser ins Sediment. Ein abfallender Sauerstoffgehalt verlängert die Pumpzeit und führt zu einer sinkenden Fließgeschwindigkeit. Außerdem wird aus den Untersuchungen eine jahreszeitliche Variabilität der Bioirrigation sichtbar, welche unabhängig von konstanten Laborbedingungen auftritt.

Mit dem Wasserstrom werden Porenwasserspezies wie SRP in den Wasserkörper abtransportiert, wohingegen SO_4^{2-} und O_2 aus dem Überstandswasser in das Sediment eingetragen werden. Ferner kommt es zur Oxidation von Fe^{2+} infolgedessen Phosphat im Sediment festgelegt wird, wie mittels P-Fraktionierung gezeigt werden konnte. Mikrobiologische Untersuchungen zeigten, dass Bioirrigation die mikrobielle Abundanz steigert, die Bakteriengemeinschaft verändert und das Potential zur enzymatischen Hydrolyse erhöht.

Chironomus plumosus, Bioirrigation, Pumprate, Fließgeschwindigkeit, Sedimentmikrobiology

Summary

The aim of this thesis is to fill gaps of knowledge regarding bioirrigation in freshwater sediments. *Chironomus plumosus* was chosen for the investigations since the filter-feeding larva dwelling in U-shaped burrows is quite common and flushes its burrow with water from the overlying water body.

To investigate bioirrigation activity in burrows of approximately 1.7 mm in diameter appropriate measurement techniques were developed. With the methods several parameters were measured for 4th stage of larvae: flow velocity (14.9 mm s^{-1}), pumping time (33 min h^{-1}), and pumping rate (61 ml h^{-1}). Consequently, a water volume equivalent to the volume of Lake Müggelsee in Berlin is pumped through the sediment every 5 days by a population density of 745 larvae m^{-2}.

The nuclear medicine imaging technique Positron Emissions Tomography was adapted and used to analyze the transport in the sediment pore water. By means of the experiments, it could be shown that even in muddy lake sediments advection is a relevant transport process and should not be neglected.

Rising temperatures result in increased pumping rates and increased influx rates of surface water into the sediment due to increased flow velocities in the burrows. Dropping oxygen concentrations prolong the pumping duration while the flow velocity decreases. Furthermore, experiments show a seasonal variability of bioirrigation which is independent of constant laboratory conditions.

Pore water species such as SRP are transported with the water flow into the overlying water body, whereas SO_4^{2-} and O_2 are transported from the overlying water into the sediment. Due to the oxidation of Fe^{2+}, phosphorus is fixed into the sediment, a result confirmed by P-fractionation. Microbiological investigations of the burrow walls demonstrated that the bioirrigation activity enhances the microbial abundance, changes the community structure, and increases the potential of enzymatic hydrolysis.

Chironomus plumosus, bioirrigation, pumping rate, flow velocity, sediment microbiology

Table of contents

Preface *3*

Figures *6*

Tables *11*

I. Introduction 13
I.I. Macrozoobenthos in aquatic sediments *13*
I.II. Bioirrigation caused hydrodynamic fluxes and biogeochemical consequences *15*
I.III. Focus of former studies *18*
I.IV. Objectives of the study and hypotheses *21*

II. Hydrodynamic transport processes 25
II.III. Bioirrigation by Chironomus plumosus: advective flow investigated by particle image velocimetry *25*
II.IV. Identification of transport processes in bioirrigated muddy sediments by [18F]fluoride PET (Positron Emission Tomography) *25*
II.V. Alteration of Chironomus plumosus ventilation activity and bioirrigation-mediated benthic fluxesby changes in temperature, oxygen concentration, and seasonal variations *25*

III. Biogeochemical impacts 26
III.I. Impacts on pore water composition *26*
III.II. Impacts on sediment composition *33*
III.III. Impacts on sediment microbiology *40*

IV. Synopsis 48
IV.I. Summary of the results *48*
IV.II. Conclusions *51*
IV.III. Outlook *56*

References *61*

Appendix 80

Acknowledgements *85*

Supplement *86*

Preface

This dissertation contains five research articles related to the investigation of the influence of macrozoobenthos on hydrodynamic transport in lake sediments. An introduction to the scientific background, a state of literature, and the objectives of this study are given in chapter I. Chapter II presents the five articles that have been published or submitted to international journals listed in the Scientific Citation Index. Furthermore, in chapter III three subchapters are composed, dealing with investigations about biogeochemical impacts of macrozoobenthos. Those results are not published in a journal by now, but are presented in several conference contributions. The most important findings of the study are compiled in chapter IV. A brief description of the five articles including my personal contributions listed below. The first author is always the corresponding author.

Chapter II.I: Measurement techniques for quantification of pumping activity of invertebrates in small burrows
Andrea Roskosch, Michael Hupfer, Gunnar Nützmann, Jörg Lewandowski
Fundamental and Applied Limnology, 178(2): 89-110 (2011)
The article presents different measurement techniques for the quantification of pumping activity in small (≤ 2 mm) macrozoobenthos burrows. The results of the pumping activity of *Chironomus plumosus* larvae were compared among the tested techniques and with data published in previous studies. All measurements, with the exception of PIV (see chapter II.II), were done at the IGB. I contributed to the paper with sampling, literature research, optimizing appropriate measurement techniques, preparation of the laboratory experiments, laboratory measurements, data interpretation, and writing the manuscript. Estimated overall contribution: 80 %.

Chapter II.II: Quantification of pumping rate of *Chironomus plumosus* larvae in natural burrows
Mohammad Reza Morad, Arzhang Khalili, Andrea Roskosch, Jörg Lewandowski
Aquatic Ecology 44(1): 143-153 (2010)
The article presents the applicability of PIV for the quantification of pumping rates in macrozoobenthos burrows. This paper was published within cooperation of the IGB and the Max Planck Institute for Marine Microbiology (MPI) in Bremen. PIV-measurements and data analysis were carried out at the MPI. I contributed the experimental setup to this paper, and I was mainly responsible for sampling and the preparation of the laboratory

experiments. I had been involved in the laboratory measurements, data interpretation, and the writing of the manuscript. Estimated overall contribution: 35 %.

Chapter II.III: Bioirrigation by *Chironomus plumosus*: advective flow investigated by particle image velocimetry

Andrea Roskosch, Mohammad Reza Morad, Arzhang Khalili, Jörg Lewandowski

Journal of the North American Benthological Society 29(3): 789-802 (2010)

The article presents the implementation of PIV for the investigation of advective flow in burrows of *C. plumosus* larvae. This paper was published within cooperation of the IGB and the Max Planck Institute for Marine Microbiology (MPI) in Bremen. PIV-measurements and data analysis were carried out at the MPI; X-ray analysis was performed at the Federal Institute for Material Research and Testing (BAM) in Berlin. I contributed to this paper with the experimental setup, and I was mainly responsible for sampling, preparation of the laboratory experiments, data interpretation, and writing the manuscript. I was involved in the laboratory measurements. Estimated overall contribution: 75 %.

Chapter II.IV: Identification of transport processes in bioirrigated muddy sediments by [18F]fluoride PET (Positron Emission Tomography)

Andrea Roskosch, Jörg Lewandowski, Ralf Bergmann, Florian Wilke, Winfried Brenner, Ralph Buchert

Applied Radiation and Isotopes 68: 1094–1097 (2010)

The article presents the applicability of PET for the analysis of bioirrigation caused advective and diffusive transport in muddy sediments. This paper was published within cooperation of the IGB and the Department of Nuclear Medicine in the University Medical Center Hamburg-Eppendorf (UKE) as well as the Radiopharmaceutical Biology Division in the Forschungszentrum Dresden-Rossendorf (FZD). PET and data analysis were performed at UKE and FZD. My contribution to the paper was the experimental setup. I was mainly responsible for sampling, the preparation of the laboratory experiments, and writing the manuscript. I was involved in the laboratory measurements and data interpretation. Estimated overall contribution: 65 %.

Chapter II.V: Alteration of Chironomus plumosus ventilation activity and bioirrigation-mediated benthic fluxesby changes in temperature, oxygen concentration, and seasonal variations

Andrea Roskosch, Nicolas Hette, Michael Hupfer, Jörg Lewandowski

Freshwater Science: In press (2012)
The article presents the implementation of a number of techniques tested in chapter II.I and examines the influence of oxygen concentration, temperature, and season on bioirrigation activity of *C. plumosus* larvae. This paper involves investigations performed by Nicolas Hette (Technische Universität Berlin) during his Diplomthesis at the IGB (July to November 2008). All measurements were conducted at the IGB. My contribution to the paper was the experimental setup, and I was mainly responsible for the measurements of the effects of season, data interpretation, as well as writing the manuscript. I was involved in sampling and the measurements of the effects of temperature and oxygen concentration. Estimated overall contribution: 70 %.

Figures

Fig. I-1. Transport processes caused by bioirrigating macrozoobenthos inside its burrow.

Fig. II-I-1. Calibration setup for flow velocity microelectrode measurements in burrow outlets. A HPLC-pump produced constant flow velocities in the Tygon tube in which the sensor tip was placed.

Fig. II-I-2. Conductivity exchange experiment with a *Chironomus plumosus* larva dwelling in a sediment-filled Perspex tube.

Fig. II-I-3. *Chironomus plumosus* larva dwelling in a water-filled transparent Tygon tube (20 cm long).

Fig. II-I-4. Flow velocity microelectrode measurement of flow velocity in an outlet of a *Chironomus plumosus* burrow; start and the stop of a pumping period are marked.

Fig. II-II-1. Left image: experimental setup showing a CCD camera, a diode laser, and the sediment tank, filled partially with sediment and overlying water. The larva is indicated in the burrow. The chimney-like elevation near the burrow inlet is constructed by the burrowing larva, and is captured by the CCD camera (right image).

Fig. II-II-2. Velocity vector field (arrows) and the color map of flow magnitudes (red stands for maximum whereas dark blue represents the minimum) above a burrow inlet of setup I.

Fig. II-II-3. The contours of the velocity (color maps) and the velocity vector plots (arrows) at (a) the inlet and (b) the corresponding outlet of setup II (wire gauze U-tube filled with sediment).

Fig. II-II-4. The pulsing pattern in the velocity contours near the outlet caused by the pumping mechanism of *C. plumosus* larva.

Fig. II-II-5. The contours and vectors of the velocity at an inlet of setup III (Tygon U-tube without sediment).

Fig. II-II-6. Streamlines obtained from the measurement plane. The circle denotes the cross section of the streamlines entering the burrow with its diameter shown as dashed line.

Fig. II-III-1. Experimental setup shows the particle image velocimetry (PIV) system, a charge-coupled device (CCD) camera, laser, illuminated water section, positioning system, and the personal computer (PC) in front of the tank with a *Chironomus plumosus* burrow inside the sediment and polyamide particles in the water

Fig. II-III-2. Schematic presentation of the stream tube concept at the burrow outlet. The tube contains all streamlines within the boundary lines (bold dashed lines). The circle (radius of the given circular cross-section = r_c) denotes the 2-dimensional plane over which the vertical velocity component (v) has been integrated. The same scheme applies for a burrow inlet, except that at the inlet the velocity field is reversed.

Fig. II-III-3. Schematic sketch of a pumping larva in setup II (water-filled Tygon tubing). Larval dimensions (larval diameter DC = 1.4 mm; length of the body fragment important for one pumping movement l_p = 10.5 mm) and dimensions of the Tygon tube (diameter DT = 3.2 mm) were used to calculate theoretical volumetric flow rate (Q_{theor}) and flow velocity (v_{theor}). Hatched areas represent the calculated amount of water pumped during a single up-and-down body movement.

Fig. II-III-4. Streamlines and flow fields above an inlet in sediment only of a *Chironomus plumosus* burrow during nonpumping (A, C) and pumping (B, D). Results from a burrow inhabited by a larva (A, B) are shown in contrast to results from a burrow inhabit by a pupa (C, D). The gray scales represent the magnitude of flow velocity (v) based on a single pair of images. Note that gray scales are different for panels A, B and panels C, D.

Fig. II-III-5. X-ray picture of a burrow in a tank inhabited by a *Chironomus plumosus* larva (setup I).

Fig. II-IV-1. (A) shows the maximum intensity projection of preparation 1 at 8-10 h after administration of [18F]fluoride; (B) shows the conceptual model of the burrow including position of the larva and water flow in burrow and through burrow walls.

Fig. II-IV-2. Time curve of the apparent burrow diameter in the dynamic PET image at the inlet part and at the outlet part of the chironomid burrow. (A) shows preparation 1; (B) shows preparation 2.

Fig. II-IV-3. Time curve of the difference of the apparent burrow diameter between outlet and inlet part of the chironomid burrow. (A) shows preparation 1; (B) shows preparation 2.

Fig. II-V-1. Schematic diagram of the measurement techniques used for the 4variables (experiment 1: pumping frequency, length, and duration;experiment 2:flow velocity;experiment 3:pumping rate; and experiment 4a, b:influx rate), and environmental variables that were manipulated in each experiment.

Fig. II-V-2. Box-and-whisker plots for pumping frequency (A), pumping length (B), and pumping duration (C) of *Chironomusplumosus* larvae at 0 to 3% (low), >3 to 12% (medium), >12 to 100% (high) O_2saturations at 10 and 20°C. Dotted linesindicate means, solid lines indicate medians, box ends are quartiles, whiskers show 2D plots, and * indicates outliers). p-valuesindicate probabilities associated with 1-way analysis of variance done at each temperature (10 and 20 °C) with O_2 group as the main effect (see Table 1 for post hoc comparisons of means and results of tests for differences between temperatures). Bars with the same letters are not significantly different.

Fig. II-V-3. Normal pumping athigh O_2 saturation (>20 %) and minor pumping atlowO_2 saturation (< 2 %) measured by an O_2microsensor in a burrow inlet (2 mm deep) in experiment 1. Data were taken from separate experimental trials. Time is coded as days, hours, and minutes.

Fig. II-V-4. Box-and-whisker plot for flow velocity of *Chironomusplumosus* larvae at 3 to 12% (low), >12 to 30% (medium), and>30 to100% (high)O_2 saturations measuredat 20 °C in experiment 2. Dotted lines indicate means, solid lines indicate medians, box ends are quartiles, whiskers show 2D plots, and * indicates outliers). Data were analysed with a Kruskal–Wallis H-test (see Table 1 for p-values). Bars with the same lower case letter are not significantly different.

Fig. II-V-5. Box-and-whisker plots for pumping frequency (A), pumping length (B), pumping duration (C), flow velocity (D), and pumping rate (E) of *Chironomusplumosus* larvae at 10 and 20 °C under high O_2 concentrations (>>50%) in experiments 1 to 3. Dotted lines indicate means, solid lines indicate medians, box ends are quartiles, whiskers show 2D plots, and * indicates outliers). Data were analysed with a t-tests. Bars with the same lower case letter are not significantly different.

Fig. II-V-6. Box-and-whisker plots for rate of advective and diffusive water influx into the sediment per individual caused by pumping of *Chironomus plumosus* larvae at the individual (1 larva = 360 larvae/m^2; experiment 4a) (A) and population level (10 larvae = 3600 larvae/m^2; experiment 4b) (B) at 10 and 20 °C. Experiment 4a wasdone in summer and autumn 2008, whereas experiment 4b wasdone in summer and autumn 2008, 2009, and 2010 (July to November each year). Dotted lines indicate means, solid lines indicate medians, box ends are quartiles, whiskers show 2D plots, and * indicates outliers). Data were analysed with t-tests. Bars with the same lower case letter are not significantly different.

Fig. II-V-7. Mean (±1 SD) rate of advective and diffusive water influx into the sediment caused by the pumping of a *Chironomusplumosus* population (10 larvae = 3600 larvae/m^2) at 10 and 20 °C during the 3-y study (experiment 4b). Experiments in grey boxes were conducted in summer and autumn 2008, 2009, and 2010 and were used to compare the rate of water influx into the sediment caused by pumping of *C. plumosus* larvae at 10 and 20 °C.

Fig. III-1. 2D peeper with a polysulfone membrane after incubation in a mesocosm, the sediment-water interface and the burrow lining of a *C. plumosus* larva are visible (brownish discolorations); screw positions are shown up.

Figs. III-2 and III-3. Two different diagrams of 2D peepers with one *C. plumosus* larva dwelling inside showing pore water concentrations of 2a) ammonium (NH_4-N), 2b) and 3b) soluble reactive phosphorus (SRP), 2c) iron(II) (Fe^{2+}), and 3c) sulphate (SO_4^{2-}). Courses of the burrow linings are marked. 3a) digital photo of one of the mesocosms with a 2D peeper inside. Depth 0 cm is presenting the sediment-water interface. The white dotted squares (2b and 3b) showing the areas of the representative chambers from which the concentrations of the non-affected sediment and the burrow lings were calculated.

Fig. III-4. Organic content determined from material of the non-affected sediment, the burrow walls and the sediment surface by loss on ignition.

Fig. III-5. Total phosphorus (TP) determined from material of the non-affected sediment, the burrow walls and the sediment surface.

Fig. III-6. P-fractionation of P-forms named by the extractant and determined from material of the non-affected sediment, the burrow walls and the sediment surface.

Fig. III-7. Concentration of a) iron (Fe^{2+}) and b) manganese (Mn^{2+}) determined from BD- and HCl-fractions in the non-affected sediment, the burrow walls, and the sediment surface.

Fig. III-8. Correlation of total phosphorus (TP) and iron (Fe^{2+}) determined from BD- and HCl-fractions.

Fig. III-9. Frozen sediment core fro the sampling of the microbial analyses.

Fig. III-9. Frozen sediment core fro the sampling of the microbial analyses.

Fig. III-10. Total bacterial abundance (mean ± min/ max) estimated by the DAPI counting method.

Fig. III-11. Bacterial abundance (mean ± min/ max) of live as well as of dead bacteria estimated by the Live/Dead *Back*Light Kit.

Fig. III-12. NMS-analysis (non metric scaling) for cDNA (active bacteria communities) and DNA (present bacteria communities) for a) Eubacteria and b) Archaea extracted for material from sediment surface, burrow walls and non-affected sediment (n=2 each).

Fig. III-13. Extra cellular enzymatic activities determined by the activities of general hydrolases (lipase, protease, and esterase) (mean ± min/ max).

Fig. III-14. Extra cellular enzymatic activities determined by the activities of phosphatases (mean ± min/ max).

Fig. IV-1. Transport processes and changes in the sediment-water chemistry and microbiology (bacterial abundance and enzymatic activity) caused by a bioirrigating *C. plumosus* larva inside its burrow.

Tables

Tab. I-1. Table of a number of six typical marine and limnic bioirrigating species including the parameters population density (individuals m^{-2}), body length (cm), burrow diameter (cm), burrow depth (cm), and burrow shape (cm). Maximum values are presented (Krüger, 1971; Stamhuis & Videler, 1998; Kristensen, 2001; Osovitz & Julian, 2002; Riisgard, 2007; Gallon et al., 2008; Edwards et al., 2009).

Table II-I-1. Techniques used in previous studies for quantifying pumping activity in burrows of bioirrigating invertebrates sorted by smallest tested burrow diameter and tube type. The tube type "sediment only" means that measurements took place in burrows directly build in sediment.

Table II-I-2. Results of all seven measurement techniques (means ± SD) are shown. The arithmetic mean of all setups (penultimate row) includes all results, while the arithmetic mean of the most reliable data (last row) excludes unreliable data sets (given in brackets). Data of PIV previously published in Morad et al. (2010) and Roskosch et al. (2010b).

Table II-I-3. Statistical results for the comparison of the parameters measured with (1) Video analyses, (2) Color tracers, (3) O$_2$ microelectrodes, (4) Flow velocity microelectrodes, (5) Thermal flow sensor, (6) PIV; compared methods are listed in table. Since normal distribution was not given, non parametric tests were chosen (Kruskal-Wallis *H*-test, Mann-Whitney *U*-test). For the *U*-test, only significant differences are mentioned.

Table II-I-4. Suitability (+ / -) and simplicity (*) of the seven different measurement techniques for (a) flow velocities, (b) pumping period frequency, (c) pumping period duration, (d) individual pumping time and the (e) individual pumping rate of *Chironomus plumosus* larvae. It should be noted that (1) video analyses are only suitable for artificial impermeable water-filled tubes. The last two columns are recommendations for combinations (i) to verify the results and (ii) to calculate the pumping rate.

Table II-II-1. Three different experimental setups used for the flow visualizations.

Table II-II-2. Time averaged maximum velocity measurements ($\bar{v}_{max}$) by PIV in all setups with the corresponding standard deviations at burrow inlets during pumping periods; $\sigma_{v_{max}}$: standard deviation in the maximum velocity, n_i: number of PIV-image pairs.

Table II-II-3. Time averaged volumetric flow rates ($\bar{Q}$) and corresponding average pumping velocities ($\bar{v}_{av}$) for all setups. σ_Q, d_b and n_i denote the standard deviation in the volumetric flow rate, burrow diameter and the number of the PIV-image pairs, respectively.

Table II-III-1. Mean (±1 SD) volumetric flow rates (Q) and flow velocities (v) based on particle image velocimetry (PIV) and color tracer measurements made above burrow inlets and outlets. In PIV, n is the number of image pairs analyzed during 1 pumping period at 1 burrow opening, and each row represents measurements at a single burrow opening. In color tracer measurements, n is the total number of measurements. For calculations of Q and v, we assumed that burrow diameter = 1.7 mm in setups I, III, IV and V, but in setup II, we used the diameter of the Tygon tube (3.2 mm). Burrow length was fixed (25 cm), except for setup I where the burrow length was determined from X-ray analysis. Measurements were done in summer at 23°C.

Table II-V-1. p-values for pairwise comparisons of pumping frequency, pumping length, and pumping duration among low, medium, and high O_2 saturation groups at 10 and 20°C (experiment 1) and for flow velocity among low, medium, and high O_2 saturation groups at 20°C (experiment 2). Superscrips indicate statistical test used to compare groups ([a] indicates analysis of variance, [b] indicates t-test, [c] indicates Mann–Whitney U test). Bonferroni-corrected p-value for significance all tests comparing a pair of groups was $p \leq 0.0167$.

Table II-V-2. The temperature coefficient (Q_{10}) for pumping frequency, pumping length, pumping duration, flow velocity, pumping rate, and rate of water influx at 10 and 20°C.

Table II-V-3. Mean (±1 SD; median in parentheses) spring/summer (April–July) and autumn September–October) advective and diffusive water influx (mL/h) into the sediment measured for 10 larvae at 10 and 20°C.

Table III-1. Concentrations (mean ± SD) of the pore water species phosphorus (SRP), ammonium (NH_4-N), iron(II) (Fe^{2+}), and sulphate (SO_4^{2-}) measured with 2D peepers in overlying water, sediment pore water, non-affected sediment, and burrow linings.

Table III-2. Results of the elementary analysis of N, C, S, and H determined from material of the non-affected sediment, the burrow walls and the sediment surface.

I. Introduction

I.I. Macrozoobenthos in aquatic sediments

Naturally, nutrient cycles in waters are balanced and in the sediment of rivers, lakes and oceans, early diagenetic processes are taking place. The life in the water is dependent on limiting elements such as phosphorus which is essential for all life on earth. Sometimes, e. g. if an overplus of phosphorus in lakes causes eutrophication and implicates algae blooms; the processes in the aquatic environment are disordered. Especially in such a case it is essential to understand the hydrodynamic transport processes and the biogeochemical turnover occurring in the water and, particularly, in the sediment. Furthermore it is important to know more about the interactions between chemical transactions and organisms living in the aquatic environment. Despite much effort to understand the processes and interactions in sediments as well as at the sediment-water interface, the knowledge is still fragmentary.

However, it is generally accepted that macrozoobenthic invertebrates have great impacts on aquatic ecosystems. By their various moving and feeding activities, hydrodynamic transport processes in the sediment are influenced and biogeochemical impacts on the sediment-water interface are obvious. Some species are just crawling through the sediment; others are building burrows with one, two or more openings that penetrate more than 2 m deep into the anoxic sediment layers (bioturbation) (Weaver & Schultheiss, 1983; Shull, 2001). Due to the absence of oxygen in the sediment, most of the tube-dwelling species induce passive or active water flow in the sediments by flushing the burrow lumens with oxygen rich water from the overlying water body (bioirrigation) (Gust & Harrison, 1981; Riisgard & Larsen, 2005). The input of oxygen causes an oxygenation of the burrow walls, and thus the sediment-water interface is considerably enhanced in bioirrigated sediments (Graneli, 1979a). Many species are performing bioirrigation not only to receive oxygen for respiration but also for the supply of food. By pumping water, phytoplankton and dissolved nutrients are transported in and metabolites and particulates are removed from the burrows and the adjacent sediment and are transported into the overlying water (Aller, 1978; Aller & Aller, 1992). The solute transport in sediments is highly important for diagenetic effects (Wang & Van Cappellen, 1996; Huettel et al., 2003).

Additionally, invertebrates are influencing solute transport, nutrient distribution, sediment structure, and microbial communities with anymore activities such as the resuspension of particulates, the secretion of silk-like material to stabilize the burrow walls or to build nets for filter-feeding (Leuchs & Neumann, 1990), and the ingestion of bacteria, phytoplankton and particles. The ingestion process is including digestion, excretion, and defecation of the detritus (Johnson et al., 1989).

The aforementioned activities are highly important for aquatic ecosystems since high densities and varieties of invertebrates are living in the sediments. As important bioirrigating species a number of marine invertebrates such as the worms *Nereis* sp., *Schizocardium sp., Arenicola marina, Maxmuelleria lankesteri* or *Urechis caupo* are well studied in literature. Moreover, a number of burrowing marine crabs and shrimps such as the shrimps *Callianassa subterranea* or *Alpheus mackayi,* and the crabs *Uca minax, Sesarma reticulatum* or *Eurytium limosum* are investigated in several studies. Larvae of Ephemeroptera or Diptera such as *Hexagenia limbata, Ephoron virgo, Sialis velata, Chironomus plumosus, C. anthracinus* or *C. thummi,* and the amphipodes *Corophium volutator and Leptocheirus plumulosus* are limnic species performing bioirrigation in freshwater sediments.

Marine or costal bioirrigating species may reach some rare to some 1,000 individuals of the same species per m^2 sediment (Riisgard & Banta, 1998; Kristensen, 2001). By contrast, in freshwater sediments some hundred to 100,000 individuals are burrowing per m^2 (McLachlan, 1977; Gallepp, 1979; Andersen & Jensen, 1991; Helson et al., 2006). Densities in freshwater sediments are often higher because limnic organisms are mostly smaller than their marine counterparts. Hence, the larger marine organisms need burrows with a wider diameter than the smaller limnic species. Some examples of the different dimensions of population density, body length, burrow diameter, and burrow depth of marine or costal and limnic bioirrigating species are given in Table I-1.

Tabelle I-1: **Table of a number of eight typical marine and limnic bioirrigating species including the parameters population density (individuals m-2), body length (cm), burrow diameter (cm), burrow depth (cm), and burrow shape (cm). Maximum values are presented (Krüger, 1971; Stamhuis & Videler, 1998; Kristensen, 2001; Osovitz & Julian, 2002; Riisgard, 2007; Gallon et al., 2008; Edwards et al., 2009).**

Environ-ment	Species	Population density (individuals m^{-2})	Body length (cm)	Burrow diameter (cm)	Burrow depth (cm)	Burrow shape
marine/	*Urechis caupo*	60	17	2.5	40	U

costal	Arenicola marina	50	20	0.8	40	J, U
	Nereis diversicolor	3,000-4,000	15	0.5	20	U, Y
	Callianassa subterranea	50	4	1	80	branched
limnic	Hexagenia limbata	500	2-3	0.5	6	U
	Sialis velata	50	1.5	0.3	2	U
	Chironomus plumosus	500-1,000	2	0.2	20	U
	Corophium volutator	100,000	1	0.3	7	U

I.II. Bioirrigation caused hydrodynamic fluxes and biogeochemical consequences

The bioirrigation activity of tube-dwelling invertebrates implies a number of hydrodynamic fluxes at the sediment-water interface as well as in the sediment pore water. Most species are pumping actively water through their burrows by performing undulating movements or peristaltic contractions of their bodies, or by using pleopods (Stamhuis & Videler, 1998; Osovitz & Julian, 2002). In some cases the pumping activity is performed more or less constantly interrupted by short and regular pauses, other species just ventilate their burrows for a few minutes once per hour. Length and frequency of the periods of active pumping and the pumped flow volume is depending on feeding behavior and body size of the organism, or external conditions such as temperature and oxygen concentration of the overlying water (Leuchs, 1986; Stamhuis et al., 1996; Osovitz & Julian, 2002). Sometimes organisms are inducing a passive water flow, because they are crawling and feeding inside their burrows. However, the pumping of water causes an advective flow through a burrow and water may penetrate through the burrow walls into the sediment. The total water volume pumped by an individual may change with species and environmental conditions, but worldwide huge volumes of water are pumped through the sediments. Thus, bioirrigation drastically increase water fluxes at the sediment-water interface and the exchange of particles and solutes is considerably enhanced.

Customarily, diffusion is known as the major transport mechanism for solutes in sediments. When macrozoobenthos is living in the sediment, bioturbation and bioirrigation as further important transport mechanisms are recognized (Aller, 1983; Van Rees et al., 1996). Since the sediment-water interface is enhanced by macrozoobenthos burrows (Graneli, 1979a), the diffusive transport is increased due to concentration gradients between the overlying water, which is pumped through the burrows, and the sediment pore water surrounding the burrows.

Besides, the pumping activity enhances the exchange of the water flowing through the burrows and the surrounding sediment pore water by advective transport. Bioirrigation caused advective fluxes are depending on the bioirrigation of the organism (pumping mechanism, flow velocity, burrow structure), the sediment permeability and the porosity of the burrow walls (Meysman et al., 2006a). To sum up, the hydrodynamic fluxes and the transport processes caused by bioirrigation are presented in Fig. I-1.

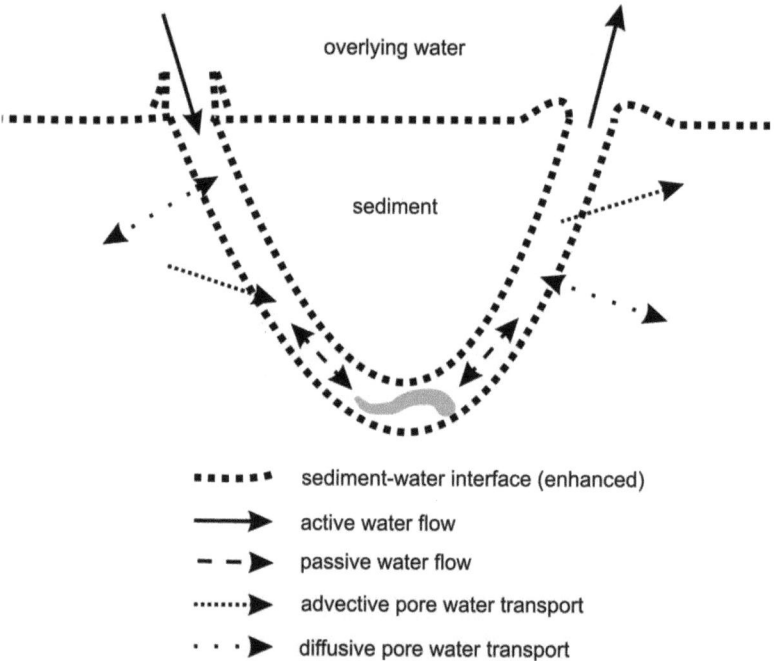

Fig. I-1: Transport processes caused by bioirrigating macrozoobenthos inside its burrow.

From literature it is well known that bioirrigation caused hydrodynamic fluxes are altering the redox potential in the sediment (Aller & Aller, 1992; Francois et al., 2002). Changed redox potentials have many consequences for biogeochemical turnover and nutrient cycling in aquatic ecosystem. Previous studies mainly investigated the impact of bioirrigation on the cycling of carbon (C) and nitrogen (N) (Kristensen, 2000; Schlüter et al., 2000; Stief & de Beer, 2006; Na et al., 2008). Some authors detected that the mineralization of organic matter and the excess of CO_2 to the overlying water body are stimulated when invertebrates periodically ventilate their burrows as oxygen is introduced into the anaerobe sediment and metabolites are removed. The removal of metabolites causes a reduced diffusion scale and enhances the total sediment metabolism. The introduction of oxygen as well as the transport of sediment into oxic environments increases the carbon oxidation of anaerobically refractory organic matter (Kristensen & Aller, 1991; Aller, 1994). Ammonification and nitrification are also stimulated by the periodically aeration of the burrows, as well as by the excretion of ammonium (NH_4^+) by the invertebrates. Denitrification is promoted because nitrate (NO_3^-) is transported from the overlying water into the adjacent sediment which is periodically anoxic (Mayer et al., 1995; Svensson & Leonardson, 1996; D´Andrea & Lopez, 1997).

Less attention has been paid to the impacts of macrozoobenthos on the cycling of phosphorous (P). Especially in limnic environments bioirrigation induced P fluxes are rarely investigated (Heilskov & Holmer, 2001), although P is known as the limiting factor responsible for the primary production of biomass and the eutrophication of lakes. The role of sediments for the turnover of P is being studied since many years (Mortimer, 1941). Usually it can be assumed that oxic sediments retain P because of the coprecipitation with oxidized iron (Fe^{3+}) and manganese (Mn^{3+}) species, while anoxic sediments release P by reduction of iron (Fe^{2+}) and manganese (Mn^{2+}) and dissolution of the complexes. Thus, the sediment is both, a source and a pool for P, whereas the oxic sediment surface acts like a diffusion barrier for P as well as for Fe^{3+} and Mn^{3+}. However, the diagenetic processes in sediments are very complex and the retention and release of P may be influenced and decoupled in various ways (Hupfer et al., 1995; Lewandowski & Hupfer, 2005b).

Besides, there is a controversy in literature if P release from or P retention in the sediment is promoted when macrozoobenthos is dwelling inside (Davis et al., 1975; Graneli, 1979b; Bostrom et al., 1988; Kalson et al., 2007; Chaffin & Kane, 2010). On the one hand, P retention might be stimulated since the oxic sediment layer is increased; on the other hand, the macrozoobenthos is responsible for increased water fluxes from the sediment into the overlying water. Moreover, bioturbation may induce P release since the oxic sediment layer is penetrated (Graneli, 1979b). The cycling of sulphur (S) is closely coupled to the P- and Fe-cycles and also an important factor for P release. If in the anoxic sediment SO_4^{2-} is reduced to hydrogen sulphide (H_2S), it competes with P for Fe^{2+} and can be precipitated as iron(II) sulphide (FeS/ FeS_2). SO_4^{2-} reduction is decreased and H_2S oxidation is increased in bioirrigated sediments (Lewandowski & Hupfer, 2005a). Consequently, less FeS is precipitated in sediments with macrozoobenthos in comparison to sediments where no macrozoobenthos is dwelling inside.

However, fluxes of solutes and particles caused by macrozoobenthos activities affect the habitat of the sediment microorganisms seriously due to the supply of electron acceptors such as oxygen and organic matter (Rosenberg, 2001; Papaspyrou et al., 2006; Mermillod-Blondin et al., 2008). Since microorganisms are known as key factors of biogeochemical processes in sediments (Battin et al., 2008) the effects of bioirrigation on microbial community structures and its activities are highly important. Although in a number of studies bioirrigated sediments and the burrow walls of tube-dwelling species show higher abundance and activity of microorganism than the non-affected sediment (Kristensen et al., 1985; Mayer et al., 1995; Papaspyrou et al., 2005), most of the effected microbial species are not identified and the parameters affecting microbial turnover rates are not completely understood.

I.III. Focus of former studies

Most studies of bioirrigation-related hydrodynamic effects focus on costal or marine tube-dwelling macrozoobenthos species (Jorgensen et al., 1986; Stamhuis & Videler, 1998; Osovitz & Julian, 2002; Riisgard & Larsen, 2005). This focus can be explained to some extend by the fact that those are usually larger than limnic species, and hence, their burrow diameters are relatively wide (compare Table I-1). Consequently, the former are better subjects for scientific studies, and most of the available measurement techniques, especially for measuring the hydrodynamic fluxes such as flow velocity and pumping rate, are appropriate for relatively wide burrow diameters. For instance, in former studies the pumping activity of marine worms, shrimps and crabs such as *Nereis* spp., *A. marina*, *U. caupo*, *Corphium* spp. or *Lembos websteri* is determined with electromagnetic flow meters, sensitive pressure measuring devices, acoustic Doppler flow probes, calorimeter or thermal anemometer (Krüger, 1964; Foster-Smith & Shillaker, 1977; Kristensen, 1981; Grove et al., 2000; Osovitz & Julian, 2002). For these techniques, burrow diameters wider than 5 mm in diameter are needed, and the organisms have to dwell in artificial tubes or in stable sediment burrows. For relatively narrow burrows, merely a flow micro sensor technique to determine low flow velocities caused by bioirrigation is suggested in literature, but was not yet tested (Riisgard & Larsen, 2001; Brand et al., 2007). Dye tracers are also used to measure flow velocities, but here the burrow length must be known (Kißner et al., 2004; Gallon et al., 2008). In several studies bioirrigation activity in a high resolution is characterized with oxygen micro sensors or planar oxygen optodes (Forster & Graf, 1995; Wang et al., 2001; Polerecky et al., 2005).

Studies dealing with the effects of changing external conditions such as season or rising temperatures due to climate change are rare. Some authors investigate the impacts of oxygen concentrations and temperature on the pumping activity of invertebrates (Walshe, 1948; Seymour, 1972; Kristensen, 1983a; Kristensen, 1983b; Leuchs, 1986; Hamburger et al., 1995); studies analyzing seasonal trends are even scarcer (Martin & Sayles, 1987; Schlüter et al., 2000). However, for the identification of essential fluxes across the sediment-water interface and for diagenetic modeling (Boudreau & Marinelli, 1994; Wang & Van Cappellen, 1996; Schlüter et al., 2000) it is essential to determine bioirrigation over a range of environmental conditions.

In literature it is often assumed that in muddy sediments the bioirrigation-mediated exchange between overlying water and sediment pore water is driven by diffusive transport. Moreover, there is some controversy about the porosity of the burrow walls. Consequently, in most studies advective transport is only be regarded in sandy sediments and is neglected for muddy environments (Huettel & Webster, 2001; Meysman et al., 2006a; Meysman et al., 2006b). Measurement techniques that allow a direct determination of advective transport in sediment pore water are rare in literature. Usually bioirrigation-mediated fluxes are examined with mathematical models (Aller & Aller, 1992; Koretsky et al., 2002; Timmermann et al., 2002; Meile et al., 2003; Meysman et al., 2006a). In experiments, the sum of diffusion and advection is determined via exchange and tracer studies, but generally these have a spatiotemporal inadequate resolution. In a number of studies the solute transport in sediments is described by the use of dissolved Br$^-$ (Aller & Aller, 1992; Rasmussen et al., 1998; Timmermann et al., 2002). NO_3^- is also appropriate to measure water exchange between pore water and the overlying water column, but can only be used as an inert tracer if the carbon content is low and the microbial activity is negligible in the sediment (Meysman et al., 2006b). In other cases, water exchange rates in sediments are calculated by lithium release (Edwards & Rolley, 1965; Fischer, 1982). Furthermore, radioactive tracers such as Radon-222 (^{222}Rn) are appropriate to analyze the pore water solute transport in sediments (Benoit et al., 1991). The application of color or fluorescent tracers such as Rhodamine in the opaque sediment is rough because for the visual detection of the tracer, the sediment has to be cut into thin slices (Precht et al., 2004). A non-invasive technique used for flow-visualization and highly resolved quantitative measurements of scalar transport through natural sediments is the Positron Emissions Tomography (PET) (Khalili et al., 1998; Khalili et al., 2000). However, this method is not yet tested for transport processes induced by bioirrigation.

The investigation of the effects of bioirrigation on the sediment biogeochemistry is often realized via pore water analysis. The sediment pore water is a key to understand biogeochemical processes and solute fluxes at the sediment-water interface. In standard procedures, samples of sediment pore water are centrifuged or squeezed. Customarily the whole sediment sample or horizontal slices of the sediment are pooled. However, these techniques affect the pore water composition, the vertical resolution is low, and horizontal gradients usually are ignored. Consequently, *in situ* sampling techniques such as diffusive dialysis samplers (Hesslein, 1976; Teasdale et al., 2010) or suction probes (Berg & McGlathery, 2001; Seeberg-Elverfeldt et al., 2005) are preferable. With these techniques, the pore water is conserved and significant chemical gradients in the sediment, both in vertically and horizontally direction can be detected. Since high-resolution sampling is necessary to investigate the influence of bioirrigation on the sediment, conventional dialysis samplers are not always suitable. 2D peepers (Lewandowski & Hupfer, 2005a; Laskov et al., 2006) and gel samplers (Krom et al., 1994; Waldbusser & Marinelli, 2006) are more appropriate to investigate the small-scale heterogeneity of pore water species. Usually the sampled pore water is analyzed by conventional methods (Zwirnmann et al., 1999; Fachgruppe Wasserchemie in der Gesellschaft Deutscher Chemiker in Gemeinschaft mit dem Normenausschuß Wasserwesen (NAW) im Deutschen Institut für Normung e.V., 1999). Since the sample volume of the aforementioned methods is very small, the pore water can be analyzed with special down-scaled methods (Laskov et al., 2006). Another possibility for high-resolution analysis of O_2, NO_3^-, or SO_4^{2-} is to scan sediment profiles with *in situ* sensors such as microelectrodes (Kühl & Revsbech, 2001; Stief & de Beer, 2002; Luther III et al., 2008).

Bioirrigation affects not only the sediment pore water, but also the sediment matrix. The analysis of the sediment matrix gives information about nutrient retention and release (Volkenborn et al., 2007). As mentioned above, many studies examine the bioirrigation-mediated effects of C and N (Andersen & Kristensen, 1991; Christensen et al., 2000; Kristensen, 2001; Heilskov & Holmer, 2001; Kristensen et al., 2009), and only a few authors are dealing with P mainly (Andersen et al., 2006; Biswas et al., 2009). Usually the sediment of a core or a mesocosm is used in total or is cut into horizontal slices. The solid matter can be analyzed by conventional methods (Zwirnmann et al., 1999; Fachgruppe Wasserchemie in der Gesellschaft Deutscher Chemiker in Gemeinschaft mit dem Normenausschuß Wasserwesen (NAW) im Deutschen Institut für Normung e.V., 1999). For the identification of different P species a fractionation method for sediment is available (Psenner et al., 1988).

The effect of burrowing macrozoobenthos on the sediment microbiology is in the focus of several studies. However, mostly the effects of bioturbation and feeding (grazing) on the microbial community and abundance (Reichardt, 1988; Yeager et al., 2001) are studied. The biomass is quantified and the cell numbers are counted by DAPI (4´,6-diamidino-2-phenylindole) or epifluorescence spectroscopy (Johnson et al., 1989; van de Bund et al., 1994; Traunspurger et al., 1997). Studies investigating the effects of bioturbation including bioirrigation on microbial activities are scarce in literature (Pelegri et al., 1994; Stief et al., 2004). Microbial activities are analyzed by determining reaction rates or mineralization (Pelegri & Blackburn, 1995; Papaspyrou et al., 2006; Mermillod-Blondin et al., 2008), and sometimes diverse enzymatic activities are quantified (Reichardt, 1988; Stief, 2007). In several studies the microbial diversity is identified by using DGGE (Denaturing gradient gel electrophoresis) and fingerprint approach (Papaspyrou et al., 2005; Bertics & Ziebis, 2009). However, commonly the microbiology of sediment inhabited by macrozoobenthos is compared to macrozoobenthos-free sediment (Gilbert et al., 1998; Mermillod-Blondin et al., 2008). Reichardt (1988) or Bertics & Ziebis (2009) analyzed the bacterial diversity of macrozoobenthos burrows in relation to the sediment surface, but costal and marine sediments inhabited by worms, shrimps, and crabs are in the focus of these and most of the of the aforementioned studies.

I.IV. Objectives of the study and hypotheses

Despite the previous chapters show that there is some research concentrating on hydrodynamic transport, biogeochemical turnover and macrozoobenthic organisms dwelling in the sediment, the understanding of the processes occurring at the sediment-water interface is still fragmentary. More effort is needed to close the gaps of knowledge, especially for freshwater environments.

Today it is well known that a variety of macrozoobenthic species influence their sedimentary environment by a number of processes including bioirrigation, bioturbation, sediment resuspension, ingestion, digestion, defecation, excretion and secretion (Lewandowski & Hupfer, 2005a). Consequences of these activities of tube-dwelling invertebrates are increased hydrodynamic transport and material exchange at the sediment-water interface due to burrow ventilation, an enhanced oxic sediment area caused by the oxidation of the burrow walls with water from the overlying water body, as well as changed biogeochemical and microbial processes due to modified redox potentials in the surrounding sediment.

However, as can be seen from the previous chapter, most former survey addressing the impact of macrozoobenthos on hydrodynamic transport and biogeochemical turnover in sediments are conducted with organisms dwelling in costal and marine environments. Investigations in freshwater environments, especially in muddy sediments typical for lakes, are comparatively scarce in literature. As shown in Table I-1, limnic macrozoobenthos species are often smaller than costal or marine invertebrates. As a consequence, limnic species that are dwelling in relative narrow burrows are less studied than their bigger marine counterparts that are living in relatively large burrows. Besides, most of the available techniques suitable for determining the pumping activity of tube-dwelling invertebrates are not appropriate for small burrow diameters ($\leq$ 2 mm). Moreover, the measurement conditions are complicated due to the soft and muddy sediments (instable, opaque material), alternating pumping conditions (intermittent or periodically pumping), and relatively low flow velocities (< 20 mm s^{-1}) as well as pumping rates (< 100 ml h^{-1}). Techniques for analyzing the bioirrigation-caused advective transport in the pore water of muddy freshwater sediments are not yet tested. Despite from the aquatic environment, the pumping rates of bioirrigating species and the bioirrigation-mediated exchange between overlying water and sediment are just estimated in a number of former studies. To conclude, there is a lack of knowledge about the relevance of pumping rates of small limnic species and, therefore, this topic should be investigated in the present study.

Furthermore, usually advection is assumed as a relevant transport process just for sandy sediments (Meysman et al. 2006). In a number of studies is given that due to the permeability of sandy sediment, invertebrates can actively pump water across the burrow wall and into the sediment when ventilating their burrows (Forster-Smith, 1978). On the contrary, muddy sediments are considered as diffusion-driven and for such environments it is accepted that advection is negligible since the water will not penetrate the muddy sediment. In the present study this theory is tested and in experiments should be shown that advection is also relevant in bioirrigated muddy sediment.

In literature the alteration of bioirrigation rates due to environmental parameters such as seasonal trends are poorly investigated. Most survey presented in the previous chapter was conducted under laboratory conditions. Hence, there is almost no competent knowledge about the consequences of changing bioirrigation activities and pumping rates under changing environmental conditions and over different seasons. To gain more information, the sensitivity of bioirrigating macrozoobenthos to environmental gradients is an objective of the present study. Therefore, the modification of bioirrigation activity with changing temperature, oxygen concentration of the overlying water and season of the year is verified.

Even though phosphorus is essential for life and responsible for eutrophication and algae blooms in waters, biogeochemical survey is mostly focused on the cycling of N and S in the aquatic environment. Consequently, little is known about the effects of bioirrigation on the cycling of P. The question if bioirrigation leads to a release or retention of phosphorus in the sediment is still not answered and, consequently, is also an objective of the present study. Since most studies are contemplate the total bioirrigated sediment or just horizontal slices are made, in the present study the small scale heterogeneity of the sediment has to be analyzed. Despite in literature some studies are dealing with the effects of bioirrigation on microbial community structures and its activities in sediments, the knowledge in this field still fragmentary, too. Moreover, most survey is focusing on costal or marine systems and thus, so far not much information is available for freshwater sediments.

Due to the aforementioned knowledge deficits, the objective of this dissertation is to evaluate the effects of bioirrigation of small but highly abundant tube-dwelling macrozoobenthic species on hydrodynamic transport in muddy lake sediments. Pumping rates and pore water transport are investigated with high resolution and over different environmental gradients. Moreover, with a focus on P, the bioirrigation-mediated biogeochemical and microbial effects are studied. To sum up, in the present study the following four main hypotheses are addressed:

Hypothesis 1 - *Bioirrigating species dwelling in narrow burrows are able to cause relevant pumping rates.*

Hypothesis 2 - *Advective pore water transport is not negligible in muddy sediments.*

Hypothesis 3 - *Bioirrigation activity is highly sensitive to environmental changes.*

Hypothesis 4 - *Bioirrigation has a significant impact on the sediment biogeochemistry and leads to enhanced P retention in the sediment.*

To test these hypotheses it is necessary to establish appropriate measurement techniques and to determine a variety of parameters. Therefore, the main objectives of the investigations are as follows:

- Select and test appropriate techniques to measure bioirrigation activity including the parameters flow velocity, frequency and length of pumping periods, individual pumping time, and individual as well as population-wide pumping rates in narrow macrozoobenthos burrows (diameter ≤ 2 mm).

- Develop adequate setups for the measurements; or rather apply a technique that makes it possible to determine the length of the burrows directly in the opaque sediment.
- Measure the rate of water influx into the sediment caused by bioirrigation activity.
- Examine the advective transport of dissolved substances in the sediment pore water around the burrow linings caused by bioirrigation activity.
- Quantify the effects of environmental parameters such as temperature, oxygen concentration, and season on pumping activity and exchange rates.
- Analyze the impact of bioirrigation on the concentration distribution of diverse pore water species, especially of sulphate around the burrow linings in comparison to the non-affected sediment.
- Investigate the impact of bioirrigation on the sediment matrix, especially on phosphorus in the burrow walls in comparison to the non-affected sediment and the sediment surface.
- Study the microbial effects of bioirrigation around the burrow linings in comparison to the non-affected sediment and the sediment surface.

To guarantee comparability of the results, the hypotheses should be tested on a single, but typical limnic species. *C. plumosus* larvae are chosen for the experiments because they are wide spread, occurring in relatively high numbers in muddy lake sediments, building U-shaped tubes with a burrow diameter $\leq$ 2 mm, and performing bioirrigation periodically for the supply of oxygen and food. For the experiments, 4th instar larvae as well as sediment and water are sampled from the shallow eutrophic Lake Müggelsee in the south east of Berlin, Germany.

II. Hydrodynamic transport processes

II.I. Measurement techniques for quantification of pumping activity of invertebrates in small burrows

Roskosch, A.; Hupfer, M; Nützmann, G. & Lewandowski, J. (2011): Measurement techniques for quantification of pumping activity of invertebrates in small burrows. Fundamental and Applied Limnology: 178(2): 89-110

II.II. Quantification of pumping rate of Chironomus plumosus larvae in natural burrows

Morad, M. R.; Khalili, A.; Roskosch, A. & Lewandowski, J. (2010): Quantification of pumping rate of Chironomus plumosus larvae in natural burrows. Aquatic Ecology 44(1): 143-153

II.III. Bioirrigation by Chironomus plumosus: advective flow investigated by particle image velocimetry

Roskosch, A.; Morad, M.R.; Khalili, A. & Lewandowski, J. (2010): Bioirrigation by Chironomus plumosus: advective flow investigated by particle image velocimetry. Journal of the North American Benthological Society 29(3): 789-802

II.IV. Identification of transport processes in bioirrigated muddy sediments by [18F]fluoride PET (Positron Emission Tomography)

Roskosch, A.; Lewandowski, J.; Bergmann, R.; Wilke, F.; Brenner, W, & Buchert R. (2010): Identification of transport processes in bioirrigated muddy sediments by [18F]fluoride PET (Positron Emission Tomography). Applied Radiation and Isotopes 68: 1094-1097

II.V. Alteration of Chironomus plumosus ventilation activity and bioirrigation-mediated benthic fluxesby changes in temperature, oxygen concentration, and seasonal variations

Roskosch, A.; Hette, N.; Hupfer, M.; Lewandowski, J. (2012): Alteration of Chironomus plumosus ventilation activity and bioirrigation-mediated benthic fluxesby changes in temperature, oxygen concentration, and seasonal variations. Freshwater Science: in press

III. Biogeochemical impacts

III.I. Impacts on pore water composition

Introduction

In a number of studies it is shown that the concentration distribution of pore water species in bioirrigated sediments are significantly different from macrozoobenthos-free sediments (Graneli, 1979b; Lewandowski et al., 2006). Tube-dwelling macrozoobenthos flushes its burrows with water from the overlying water body and accelerates the transport of pore water species such as ammonia or phosphate out of the sediment (Matisoff et al., 1985; Svensson, 1997; Aller & Aller, 1998). Due to the redistribution of the ions, the redox potential in the sediment is changed and the biogeochemical processes, which are important for diagenesis processes in aquatic ecosystems, are seriously affected (Fischer, 1982; Kristensen, 2000).

To examine the spatial relationship between the burrow linings irrigated by *C. plumosus* larvae and the concentrations of ions in the sediment pore water as well as in the overlying water, the distribution of the pore water species phosphorus (PO_4, here: soluble reactive phosphorus (SRP)), ammonium (NH_4, here: NH_4-N), iron(II) (Fe^{2+}), and sulphate (SO_4^{2-}) are determined. So far, SO_4^{2-} is not analyzed for macrozoobenthos burrows. Investigations are performed with two dimensional pore water samplers with a high spatial resolution (2D peepers) (Lewandowski et al., 2002) and down-scaled photometric methods (Laskov et al., 2006).

Material and Methods

As described in literature (Lewandowski et al., 2002; Lewandowski et al., 2007), 2D peepers with a resolution of 9 mm (380 mm high; 243 mm wide; 23 mm thick; 550 chambers; 650 µl volume each chamber) (Fig. III-1) were filled with oxygen-free distilled water and enclosed with a polysulfone membrane (0.2 µm tick) by using screws. One peeper was placed into a mesocosm made of Perspex glass (390 mm high; 244 mm wide; 29 mm deep), directly at its back wall. Thus, only a 6 mm deep space was left between the front of the peeper and the mesocosm front wall.

Sediment, water, and *C. plumosus* larvae (4th larval stage, 20 mm body length) were sampled from Lake Müggelsee in Berlin, Germany (6 m water depth, N 52°44´ and E 13°65´). Sieved sediment (≤ 0.1 mm) was filled into a mesocosm (200 mm high) and topped with water (180 mm high). Directly after one larva was added to build its burrow inside the sediment, the mesocosm was incubated in darkness at 10 °C. The overlying water was aerated with an air pump. Two mesocosms were provided to have the capacity to measure all four parameters (double batches each). After an incubation of three (SRP, NH_4-N, and Fe^{2+}) or eight (SO_4^{2-} and SRP) weeks the sediment was sieved and the remaining larvae were counted.

SRP, NH_4-N, Fe^{2+}, and SO_4^{2-} were analyzed photometrically (Sunrise, Tecan, Switzerland) following down scaled versions (Laskov et al., 2006) of the molybdenum blue method (SRP) (Murphy & Riley, 1962), the indophenol method (NH_4-N) (Krom, 1980), the phenanthroline method (Fe^{2+}) (Tamura et al., 1974), and the barium-gelatin method (SO_4^{2-}) (Tabatabai, 1974). Immediately after the peeper was removed from the mesocosm it was cleaned from sediment with distilled water. The pore water samples were transferred rapidly from the peeper in microtitre plates by using eight-channel pipettes penetrating the polysulfone membrane. The samples were measured by using a microtitre plate reader (Sunrise, Tecan, Switzerland).

For data analysis, 2D illustrations of the pore water concentrations were prepared with the software Surfer (V 5.01, Golden Software, Inc., USA). Missing data (e.g. contaminated samples or screw positions) were interpolated by using linear kriging as the gridding method. The concentrations of the pore water species in the different areas of the mesocosm were statistical analyzed with the software SPSS (version 9.0.1, SPSS Inc., USA). To differ between the pore water around burrow linings and the pore water of the non-affected sediment, representative chambers of the 2D peepers (Fig. III-2b and 3b) were pooled. Based on these values arithmetic means and relative standard deviations (mean ± SD) were calculated. The p value was reported when the difference of the results was statistically significant ($p \leq 0.05$).

Fig. III-1. 2D peeper with a polysulfone membrane after incubation in a mesocosm, the sediment-water interface and the burrow lining of a C. plumosus larva are visible (brownish discolorations); screw positions are shown up.

Results

The course of a burrow in the sediment can be detected since the aerated and therefore brownish discolorations of the sediment were visible through the transparent mesocosm and on the 2D peeper (Fig. III-1). Hence, in the experiment burrow linings were easily distinguished from the anoxic, dark brown sediment (Fig. III-3a). All larvae were found alive after sieving the sediment of the mesocosms.

In the overlying water, the concentrations of SRP, NH_4-N, and Fe^{2+} were relatively low, and the concentration of SO_4^{2-} was relatively high in comparison to the sediment pore water (Figs. III-2a, b, and c, Figs. III-3b and c). The mean concentration of the overlying water was calculated from the first three rows of chambers of a peeper, a representative concentration of the sediment pore water was calculated below the sediment-water interface from row 6 to 23. Arithmetic means (Table III-1) demonstrate that SRP, NH_4-N, and Fe^{2+} were significantly higher (SRP and NH_4-N: U-test, $p<0.001$; Fe^{2+}: t-test, $p<0.001$) and SO_4^{2-} was significantly lower (t-test, $p<0.001$) in the sediment pore water than in the overlying water.

The interface between sediment and overlying water is visible in Fig. III-3a) (changes noticeable from chambers in row 4 to 5). From Figs. III-2a, b, c, and II-3c it can be observed that directly at the sediment-water interface the concentrations of SRP (from 47.5 to 2750 and from 85.8 to 366 µg L^{-1}, respectively), NH_4-N (from 46.4 to 378.1 µg L^{-1}), and Fe^{2+} (from 78.3 to 11,900 µg L^{-1}) were sharply increased (depth 0 cm in Figs. III-2 and III-3. Fig. III-2b shows that the concentration of SO_4^{2-} (from 217 to 218 mg L^{-1}) was not changing immediately, but was decreased down in the sediment with distance to the sediment-water interface.

The high standard deviations of the pore water concentrations demonstrated a high variability of SRP, NH_4-N, Fe^{2+}, and SO_4^{2-} in the sediment. This variability is also obvious in the diagrams of the 2D peepers (Figs. III-2 and III-3). In comparison to the non-affected sediment, relatively low concentrations of SRP, NH_4-N, and Fe^{2+}, and relatively high concentrations of SO_4^{2-} were obvious along and around the burrow linings (Fig. III-3a). Values given in Table III-1 were calculated from representative chambers of the 2D peepers (Figs. III-2b and III-3b). The chambers representing the sediment around the burrow linings were chosen from the same depth as the chambers representing the non-affected sediment. SRP, NH_4-N, and Fe^{2+} were significantly higher (t-test, $p<0.001$, for all), and SO_4^{2-} was significantly lower (t-test, $p<0.001$) in the non-affected sediment than around the burrow lings. Although the SRP concentrations were slightly increasing in the overlying water and slightly decreasing in the non-affected sediment, no significant differences between the 2D peepers incubated three and eight weeks were determined.

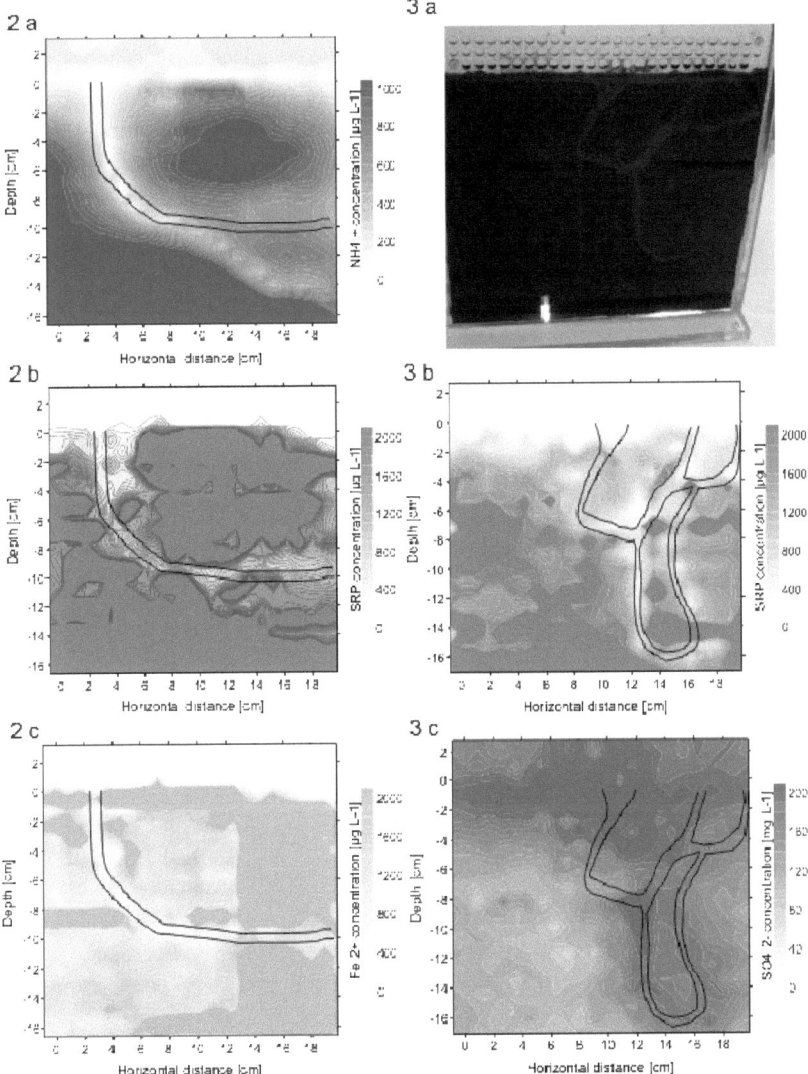

Figs. III-2 and III-3: Two different diagrams of 2D peepers with one C. plumosus larva dwelling inside showing pore water concentrations of 2a) ammonium (NH4-N), 2b) and 3b) soluble reactive phosphorus (SRP), 2c) iron(II) (Fe^{2+}), and 3c) sulphate (SO_4^{2-}). Courses of the burrow linings are marked. 3a) digital photo of one of the mesocosms with a 2D peeper inside. Depth 0 cm is presenting the sediment-water interface. The white dotted squares (2b and 3b) showing the areas of the representative chambers from which the concentrations of the non-affected sediment and the burrow lings were calculated.

Table III-1. Concentrations (mean ± SD) of the pore water species phosphorus (SRP), ammonium (NH4-N), iron(II) (Fe2+), and sulphate (SO42-) measured with 2D peepers in overlying water, sediment pore water, non-affected sediment, and burrow linings.

Mesocosm-no./ Figure	Parameter concentration	Overlying water	Pore-water	Non-affected sediment	Burrow linings
1/ III-2a	SRP ($\mu g\ L^{-1}$)	43.8 ± 9.57 n=72	2640 ± 1810 n=420	3250 ± 2120 n=36	249 ± 197 n=10
1/ III-2b	NH_4-N ($\mu g\ L^{-1}$)	158 ± 17.4 n=72	914 ± 419 n=420	1070 ± 58.4 n=36	84.5 ± 77.1 n=10
1/ III-2c	Fe^{2+} ($\mu g\ L^{-1}$)	15.1 ± 26.2 n=72	2720 ± 2730 n=420	1340 ± 344 n=36	494 ± 188 n=10
2/ III-3b	SRP ($\mu g\ L^{-1}$)	81.8 ± 18.6 n=72	1810 ± 1210 n=420	2230 ± 1380 n=43	261 ± 92.9 n=10
2/ III-3c	SO_4^{2-} ($mg\ L^{-1}$)	191 ± 19.7 n=72	147 ± 42.1 n=420	104 ± 21.9 n=43	218 ± 18.7 n=18

Discussion

Bioirrigation considerably effects the composition of the sediment pore water. The concentrations of the pore water species SRP, NH_4-N, Fe^{2+}, and SO_4^{2-} are drastically changed along the linings of *C. plumosus* burrows. SRP, NH_4-N, and Fe^{2+} are strongly decreased, and SO_4^{2-} is strongly increased in the parts of the sediment where the burrow linings are visible due to the oxidized sediment (Lewandowski & Hupfer, 2005a) in comparison to the anoxic non-affected sediment.

Decreased pore water concentrations of SRP, NH_4-N, and Fe^{2+} can be traced to the activity of a larva inside its burrow (Graneli, 1979b; Lewandowski et al., 2007). Due to the transportation of overlying water through a burrow (compare chapter II.II), the burrow linings are flushed intensively. As it can be observed from the results, in the overlying water concentrations of SRP, NH_4-N, and Fe^{2+} are significantly lower than in the non-affected sediment. In the sediment pore water concentrations of SRP, NH_4-N, and Fe^{2+} are increasing with distance to the burrow walls, and the lowest concentrations are found directly at the burrow walls. As discussed in chapter II.IV, the overlying water is transported into the sediment since advective transport is caused by pressure gradients inside the burrow. Moreover, intensive burrow-flushing increases the contact between the sediment pore water and the overlying water. Hence, the ions are transported out of the sediment pore water and through the burrow walls by concentration gradients (Aller, 2001). Finally, with the water inside burrow the ions are pumped out of the sediment and into the overlying water body. Thus, the pore water species are transported by bioirrigation-caused advective and diffusive hydrodynamic processes.

In contrast, the pore water concentration of SO_4^{2-} is increasing with vicinity to the burrow linings. The concentration of SO_4^{2-} is higher in the overlying water than in the non-affected sediment since in an anoxic environment SO_4^{2-} is reduced to sulphide (S^{2-}). SO_4^{2-} is transported from the overlying water into the burrow lumens by pumping. Finally, SO_4^{2-} is transported from the burrow lumens into the sediment pore water around the burrow linings by advective (pressure gradients) and diffusive (concentration gradients) processes. Besides, the oxidized burrow walls are causing a re-oxidation of the S^{2-} in the sediment (Joergensen & Bak, 1990).

Oxidized burrow walls and thus changed redox potentials around the burrow linings are originated from the oxygen of the overlying water, which is transported into the anoxic sediment by active pumping of the larvae (Graneli, 1979a; Polerecky et al., 2006). This influences the microbial community structure and the metabolic activity of the microorganisms in the sediment (Johnson et al., 1989; van de Bund et al., 1994; Bertics & Ziebis, 2009). Moreover, the nutrient situation is changed around the burrows because digestion and excretion of the larvae conditioned the availability of nutrients for the microorganisms (Kristensen & Mikkelsen, 2003). Besides, the secretion of the larvae that stabilizes the burrow walls leads to an altered sediment milieu. However, increased metabolic activity of the microorganisms should result in an increased uptake of PO_4^{3-} and NH_4^+, and consequently in decreased concentrations of pore water ions (Tezuka, 1990). Besides, some microorganisms are able to take up phosphate beyond their immediate physiological needs and to store it in their cells (Hupfer & Uhlmann, 1991). This process has been technically optimized for P elimination in sewage plants (Hupfer et al., 2007).

Organic P was also determined in the oxidized walls of *C. plumosus* burrows (unpublished data).

Another consequence of oxidized burrow walls is the oxidation of iron(II) (Fe^{2+}) to iron(III) (Fe^{3+}). The Fe^{3+} precipitates as Fe(III) oxyhydroxide (Fe(OOH)) and absorbs the PO_4^{3-} from the pore water (Fe(OOH)-PO_4) (Gunnars et al., 2002). Finally, in the sediment P might be fixed as ferrousphosphate ($Fe_3(PO_4)_2$) under anoxic conditions. Furthermore, due to nitrification processes in the burrow walls, NH_4^+ is oxidized into nitrite (NO_2^-) and nitrate (NO_3^-) (Stief & de Beer, 2002). Consequently, concentrations of SRP, NH_4-N, and Fe^{2+} are further decreased in the sediment around the burrow linings. However, in the reducing conditions of the anoxic sediment, Fe(OOH)-PO_4 is brought into solution and the concentrations of SRP and Fe^{2+} in the non-affected sediment are increased again (Mortimer, 1941).

When the concentrations of SRP and NH_4-N (depth -10 to -14 cm, horizontal distance 8 to 18 cm) are considered unflushed burrow linings become obvious (Figs. III-2a and III-2b). This observation implies that the resolution of Fe(OOH)-PO_4 and the additional supply of NH_4^+ are relatively slow, and some time is needed until the conditions are the same than in the non-affected sediment again. In the pore water concentration of Fe^{2+}, these old burrows are not detected since the surplus Fe^{3+} is reduced to Fe^{2+} immediately (Fig. III-2c).

In Fig. III-2c (horizontal distance 13 cm), a gradation in the concentration of Fe^{2+} is obvious between the different microtitre plates. This effect may be caused by a time delay during the measurement procedure or an error in pipetting. The values in the concerned area are not chosen as representative chambers for the calculation of the mean concentrations of the non-affected sediment.

A shortcoming of the peeper technique is that a temporal resolution is not possible since just the equilibrium of the sediment pore water at a proper time (here: three or eight weeks) is analyzed (Figs. III-2 and III-3). Other techniques such as PET (chapter II.IV) have to be used to observe a spatial-temporal distribution in the sediment. Moreover, differences between inlet and outlet side of a burrow are not detectable with the method.

To summarize, the pore water chemistry around the linings of a macrozoobenthos burrow differs significantly from the non-affected sediment. The heterogeneity of the pore water is caused by several reasons. The superior reason is the flushing of the burrow lings caused by bioirrigation. Hence, the concentrations of SRP, NH_4-N, Fe^{2+}, and SO_4 are directly affected by advective transport caused by pressure differences and diffusive transport caused by concentration gradients. Thus, an exchange between the water in a burrow and the adjacent sediment pore water is accelerated. Indirectly, the sediment around the burrow linings is affected by burrow-flushing since the burrow walls are oxidized. Therefore, the ion concentrations in the pore water are changed due to altered redox reactions and enhanced microbial processes in the burrow walls.

III.II. Impacts on sediment composition

Introduction

The effects of tube-dwelling macrozoobenthos on the pore water chemistry around the burrows described in chapter III.I must also have consequences on the composition of the sediment matrix. Mainly the effects of bioirrigation on the distribution and cycling of the nutrients carbon (C) and nitrogen (N) are studied in literature (Pelegri & Blackburn, 1996; Hansen et al., 1998; Christensen et al., 2000; Kristensen, 2001; Stief & de Beer, 2002; Stief & de Beer, 2006). This dissertation focuses on phosphorus (P) since the influence of bioirrigation on the retention or release of P is controversially discussed in literature and is still not totally clarified (Gallepp et al., 1978; Hansen et al., 1998; Lewandowski & Hupfer, 2005a). Commonly P is accepted as the limiting factor in aquatic environments, especially in freshwaters, and responsible for eutrophication and phytoplankton blooms (Schindler, 1978; Edmondson & Lehman, 1981; Karl, 2000).

In order to show the concentration differences of P in the walls of *C. plumosus* burrows in comparison to the sediment that is not affected by bioirrigation, samples of the burrow walls, the non-affected sediment as well as the oxidized sediment surface are analyzed separately. The P-species are investigated by P-fractionation (Psenner et al., 1984; van de Bund et al., 1994; Hupfer et al., 1995). Additionally, the concentration of the total P, the dry weight (DW), the loss on ignition (organic content = OC), as well as the concentrations of iron (Fe^{2+}), manganese (Mn^{2+}), carbon (C), nitrogen (N), sulphur (S), and hydrogen (H) are determined.

Material and Methods

Sieved sediment ($\leq$ 0.1 mm) from Lake Müggelsee (compare chapter III.I) was filled into mesocosms made of Perspex (200 mm high; 250 mm wide; 60 mm depth; 170 mm sediment height) and topped with water from Lake Müggelsee (25 mm high). After two *C. plumosus* larvae (4th larval stage, separated from sediments of Lake Müggelsee, compare chapter III.I) were added to build their burrows inside the sediment, the mesocosms were incubated in darkness at 10 °C (double batches). The overlying water was aerated with an air pump. After an incubation of three weeks sediment from the burrow walls (oxidized, brownish color), the non-affected sediment (anoxic, dark grey color), and the sediment surface (oxic, brownish color) were sampled. For sediment sampling a syringe was extended with a tube and material from all three areas (double samples per Perspex box) was carefully sucked up.

After sampling, DW and OC were determined conventionally, and the total P (TP) was analyzed photometrically (Zwirnmann et al., 1999). For P-fractionation, samples were analyzed by using a method according to Psenner et al. (1984) and modified by Hupfer et al. (1995) (Lewandowski, 2002). The concentrations of Fe^{2+} and Mn^{2+} were determined from the BD- and HCl-fractions (Table III-2) with atomic absorption spectroscopy (AAS 3300, Perkin Elmer, Inc., Germany) (Zwirnmann et al., 1999). For elementary analyses of N, C, S, and H the samples were analyzed with combustion analyses (Vario EL, Elemtar, Inc., Germany) (Zwirnmann et al., 1999).

Statistical analyses were performed with the software SPSS (version 9.0.1, SPSS Inc., USA). Based on the means of double samples, arithmetic means and relative standard deviations (mean ± SD) were calculated. The p value was reported when results were statistically significant ($p \leq 0.05$).

Results

The DW of the sediment surface was 6.96 ± 1.78 % (n=3), of the non-affected sediment 12.13 ± 1.95 % (n=3), and of the burrow walls 6.68 ± 2.06 % (n=5). The OC determined by the loss on ignition (Fig. III-4) was significantly not different for burrow walls (n=5; 24.2 ± 0.83 %), sediment surface (n=3; 24.4 ± 0.61 %), and non-affected sediment (n=3; 22.0 ± 2.11 %).

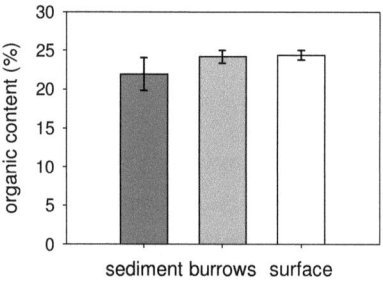

Fig. III-4. Organic content (OC) (mean ± SD) determined from material of the non-affected sediment, the burrow walls and the sediment surface by loss on ignition.

The TP concentration (Fig. III-5) was significantly higher in the burrow walls than in the non-affected sediment (U-test, $p=0.025$). The TP determined in the sediment surface was also significantly higher than in the non-affected sediment (U-test, $p=0.050$). The concentration in the burrow walls (n=5; 3.69.2 ± 0.62 mg g^{-1}) was more related to the sediment surface (n=3; 2.97 ± 0.39 mg g^{-1}) than to the non-affected sediment (n=3; 1.91 ± 0.43 mg g^{-1}), but no significant differences were observed between burrow walls and sediment surface.

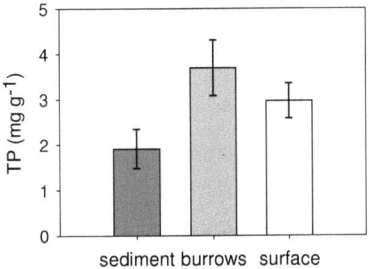

Fig. III-5. Total phosphorus (TP) (mean ± SD) determined from material of the non-affected sediment, the burrow walls and the sediment surface.

The P-fractionation (Fig. III-6) resulted in the same outcome than the TP measurements. The total amount of P calculated by adding up the concentrations of the single P-fractions was higher in the burrow walls (n=3; 3.74.2 ± 0.64 mg g^{-1}) than in the sediment surface (n=1; 2.88 mg g^{-1}), but the lowest concentration was determined in the non-affected sediment (n=1; 2.06 mg g^{-1}). In the three sample areas, the highest variability was detected for the BD-fraction (bicarbonate/ dithionite). Therefore, the organic and redox sensitive Fe^{2+}- and Mn^{2+}-bound P is the most variable P-species.

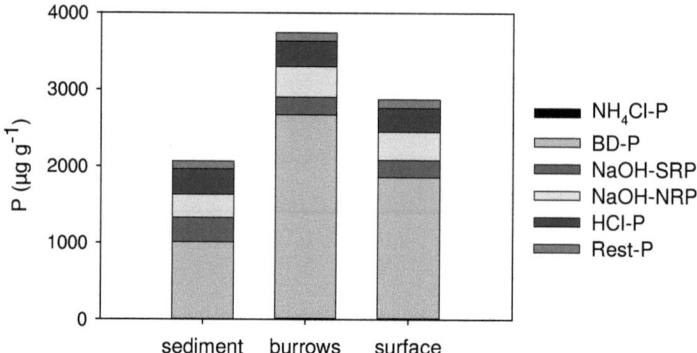

Fig. III-6. P-fractionation of P-forms named by the extractant and determined from material of the non-affected sediment, the burrow walls and the sediment surface.

The concentration of Fe^{2+} determined from BD- and HCl-fractions (Fe^{2+}_{BD+HCl}) (Fig. III-7a) resulted in slightly higher values for the burrow walls (n=3; 65.6 ± 1.60 mg g^{-1}) and the sediment surface (n=1; 64.1 mg g^{-1}) than for the non-affected sediment (n=1; 58.2 mg g^{-1}). A high correlation (r^2=0.89) between TP and Fe^{2+}_{BD+HCl} was found as demonstrated in Fig. III-8.

The concentration of Mn^{2+} determined from BD- and HCl-fractions (Mn^{2+}_{BD+HCl}) (Fig. III-7 b) was clearly higher in the sediment surface (n=1; 4.77 mg g^{-1}) than in the burrow walls (n=3; 2.05 ± 0.53 mg g^{-1}), while the lowest concentration was detected in the non-affected sediment (n=1; 0.90 mg g^{-1}).

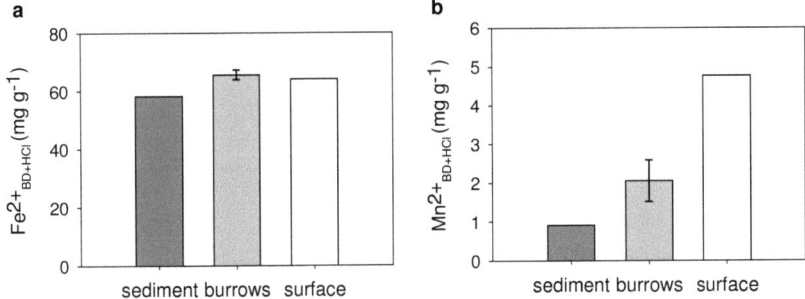

Fig. III-7. Concentration of a) iron (Fe2+) and b) manganese (Mn2+) (mean ± SD) determined from BD- and HCl-fractions in the non-affected sediment, the burrow walls, and the sediment surface.

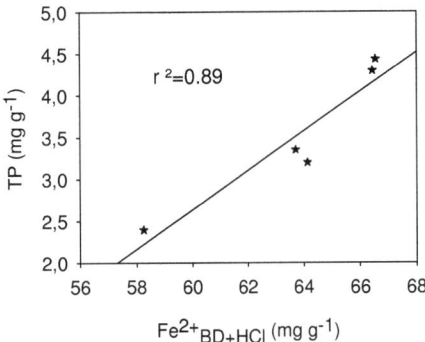

Fig. III-8. Correlation of total phosphorus (TP) and iron (Fe2+) determined from BD- and HCl-fractions.

For the concentrations of N, C, and H, the results of the non-affected sediment, the burrow walls, and the sediment surface differ not significantly. The concentration of S was significantly lower in the burrow walls (U-test, $p=0.025$) and in the sediment surface (U-test, $p=0.050$) than in the non-affected sediment. The results of the elementary analyses of C, N, S, and H are presented in Table III-2.

Table III-2. Results of the elementary analysis of N, C, S, and H determined from material of the non-affected sediment, the burrow walls and the sediment surface.

	n	N (mg g^{-1}) mean ± SD	C (mg g^{-1}) mean ± SD	S (mg g^{-1}) mean ± SD	H (mg g^{-1}) mean ± SD
sediment	3	12.0 ± 1.54	148 ±12.6	15.4 ± 1.27	14.9 ± 0.84
burrows	5	12.0 ± 1.86	149 ± 13.6	11.9 ± 0.73	15.9 ± 1.48
surface	3	12.4 ± 1.19	151 ± 10.1	12.8 ± 0.96	15.9 ± 1.64

Discussion

The composition of the sediment matrix is significantly influenced by bioirrigation. The P-concentration in the sediment burrow walls is significantly higher than in the non-affected sediment. The heterogeneity of the sediment matrix detected for TP is proven by P-fractionation. Here, most differences are obvious in the BD-fraction that includes the organic and redox sensitive Fe^{2+}- and Mn^{2+}-bound P.

The high P-concentration measured in the burrow walls as well as in the sediment surface can be explained by the oxidation of Fe^{2+} into Fe^{3+} and the sorption of P out of the sediment pore water by the formation of $Fe(OOH)-PO_4$ (Gunnars et al., 2002; Lewandowski & Hupfer, 2005a). Previously, Fe^{2+} is transported from the surrounding sediment to the burrow walls and to the sediment surface by diffusive processes (chapter III.I). Burrow walls and sediment surface are both oxidized with oxygen from the overlying water. The sediment surface is oxidized by mainly diffusive processes and the burrow walls are oxidized by bioirrigation-mediated flushing of the burrow lumens. Diffusive and advective processes (chapter III.I) transport the oxygen from the water inside the burrow into the burrow walls and the adjacent sediment. The precipitation of P incorporated in $Fe(OOH)-PO_4$ is confirmed by the P-fractionation since the Fe^{3+}-bound P-fraction (BD-fraction) is the most important and variable one. For the anoxic sediment that is not affected by bioirrigation the lowest value was determined.

The high correlation of P and Fe^{2+} (BD- and HCl-fraction) is also confirmed by the co-dependency of these nutrients. However, concentrations of Fe^{2+} are slightly higher but not significantly different in the burrow walls and the sediment surface than in the non-affected sediment. Most likely differences are not visible since in the burrow walls and the sediment surface Fe^{2+} is oxidized to Fe^{3+}, and precipitates after the formation of Fe(OOH). In the non-affected sediment Fe^{2+} is available as Fe(II) sulphide (FeS). As investigated in further studies (Lewandowski & Hupfer, 2005a), in bioirrigated sediments, SO_4^{2-} reduction is decreased, S^{2-} oxidation is increased, and the consumption of SO_4^{2-} is reduced. Consequently, less Fe^{2+} is precipitated as FeS while more Fe^{2+} is precipitated as Fe(OOH)-PO_4 in bioirrigated sediments (around the burrow linings) than in macrozoobenthos-free sediments. However, in sulphidic sediments only a little amount of the Fe^{2+} is mobile and relatively few oxidized Fe^{3+} is necessary to fix the available P concentrations. Hence, no enrichment of Fe^{2+} at the oxidized boundary layers occurred.

The concentration of S is significantly higher in the non-affected sediment than in the burrow walls and the sediment surface. This also indicates that in the oxidized sediment areas less S is precipitated than in the anoxic non-affected sediment. The elementary analyses results in no significant differences for C, N, and H.

The concentration of Mn^{2+} determined from the BD- and HCl-fractions is more than four times higher in the sediment surface than in the non-affected sediment. The concentration in the burrow walls is still twice as high as in the non-affected sediment. The P-fractionation shows that the P concentration in the HCl-fraction is relatively low and remains constant in all three sediment areas, but the concentration of P in the BD-fraction is much higher and highly variable. Therefore, the concentration differences of Mn^{2+} might indicate that at the sediment surface more P is bound to Mn^{2+} than to Fe^{2+}. It might also indicate that most of the P is precipitated as Fe(OOH)-PO_4. However, since the N is not very high, an assured statement is difficult.

The organic content is slightly lower in the non-affected sediment than in the burrow walls and the sediment surface. However, differences are not significant because the standard deviation of the non-affected sediment is relatively high. The high standard deviation of the non-affected sediment might indicate that the sediment was not homogenized enough before it was filled in the different mesocosms.

To conclude, in the material collected from the burrow walls significantly higher P-concentrations are detected than in the non-affected sediment. Consequently, the oxidized zones around the burrow linings function as boundaries for P release from the sediment similar to the oxic sediment surface. Thus, the P that is transported from the overlying water into the burrow lumen by active pumping of the larvae is fixed in the sediment. However, since most of the P is bound to Fe^{2+} and Mn^{2+}, the conditions are highly susceptible to changes in the redox potential. Long-term effects and the re-solution of the precipitated Fe(OOH)-PO$_4$ after reduction of the burrow walls caused by the disappearance of the larvae have to be analyzed in further studies.

III.III. Impacts on sediment microbiology

Introduction

The irrigation of the burrow linings and the gut of the burrowing invertebrates alter the sediment around macrozoobenthos burrows. The structure and diversity of microbial communities are changed and a variety of microbial processes in the burrow walls and the adjacent sediment are stimulated (Johnson et al., 1989; van de Bund et al., 1994; Bertics & Ziebis, 2009). Consequently, the microbial driven cycling of nutrients in aquatic ecosystems as well as the physical characteristics of the sediment, and thus the habitat of the microorganisms, is influenced by bioirrigation (Steward et al., 1996; Laverock et al., 2010). However, little is known about the distribution, diversity, and function of the microbial communities that inhabit the macrozoobenthos burrows and the adjacent sediment (Papaspyrou et al., 2006). So far, the bioirrigation caused alterations in the microbial transformations of nutrients at the sediment-water interface, which are not completely understood. Hence, the context of bioirrigated sediment and microbial activities and community structures should be clarified in more detail.

One main question discussed in literature is if oxidized burrow walls and sediment surface support the same microbial communities and activities since the environmental conditions are similar (Bertics & Ziebis, 2009). Other authors hold the view that communities and activities in the burrow walls are considerably different than in the surface (Kristensen & Kostka, 2005; Papaspyrou et al., 2006).

Samples of the three different sediment areas are analyzed separately in this study in order to identify whether the microbial communities in the oxidized walls of *C. plumosus* burrows differ from the sediment surface and the non-affected sediment or not. Subsequently, the microbial abundance is determined by counting the bacterial numbers (DAPI and Live/Dead *Back*Light Kit). DNA-sequencing including polymerase chain reaction (PCR), denaturing gradient gel electrophoresis (DGGE), and non-metric-scaling (NMS), and the measurement of enzymatic activities (general hydrolases, phosphatases) are used to analyze the microbial community structure and its activities.

Material and Methods

Sieved sediment (≤ 0.1 mm) from Lake Müggelsee (compare chapter III.I) was either filled into narrow mesocosms made of Perspex (168 mm high; 250 mm wide; 10 mm depth; 140

mm sediment height) or into cylindrical mesocosm made of steel (42 mm diameter; 278 mm high; 250 mm sediment height) and topped with water from Lake Müggelsee (25 mm high). After two *C. plumosus* larvae (4th larval stage, separated from sediments of Lake Müggelsee, compare chapter III.I) were added (double batches) to build their burrows inside the sediment, the mesocosms were incubated in darkness at 10 °C for three weeks. The overlying water was aerated with an air pump.

To determine bacterial numbers and enzymatic activities, material from the burrow walls, the non-affected sediment, and the sediment surface were sampled from the mesocosm made of Perspex with an extended syringe as described in chapter III.II and analyzed immediately. For the DNA-sequencing, the cylindrical mesocosm made of steel were inserted into a vessel filled with liquid nitrogen. After approximately five minutes, the frozen sediment cores were removed and samples (double samples) from the burrow walls, the non-affected sediment, and the sediment surface were dissected with a scalpel. The sampling was performed in a climatic chamber at 4 °C. Subsequently, these samples were frozen at -80 °C until analysis.

The total bacterial abundance was estimated by the DAPI counting method (Porter & Feig, 1980; Andersen & Kristensen, 2002), and the live fraction by using a Live/Dead *BackL*ight Kit (Molecular Probes Inc., Eugene, Oregon, USA) (Boulos et al., 1999; Haglund et al., 2003).

For the characterization of the microbial communities (analysis realized in collaboration with Claudia Dziallas, IGB), DNA and RNA of the frozen samples were extracted and a part of the 16S rRNA-gene respectively the 16S rRNA (cDNA) was amplified by using a PCR (Zhou et al., 1996). Furthermore, the communities of Eubacteria (amplified with DGGE primers 341f-GC and 907r) and Archaea (nested approach with 21f and 1492r first and then 344f-GC and 915r) were characterized by DGGE analyzes. The characterization was realized on DNA- and cDNA-level since the present (DNA) as well as the active (cDNA) communities were identified. Afterwards, cluster analysis using the dice-algorithm (comparison presence/absence of the DGGE-components) was applied to show similarities of the samples from the three different sediment areas. Therefore, NMS-analysis (non-metric-scaling) was applied to present the results.

The extra cellular enzymatic activities were identified by determining the activities of general hydrolases (lipase, protease, and esterase) and the phosphatases using diacetylfluorescein (FDA in 98 % acetone) and methylumbelliferyl phosphate (MUF-P) as model substrates, respectively (Marxsen et al., 1998; Reiche et al., 2009). In this study, the methods were modified and adapted to the Lake Müggelsee sediment. 2.5 g sediment sample were filled up with 100 g distilled water on a balance and mixed on a stirrer. During stirring, 4 ml of the suspension (triple samples) were pipetted and mixed with 1 ml of FDA (0.5 mmol L^{-1}) and MUF-P (5 mmol L^{-1}) substrate, respectively. Afterwards, the samples were incubated on an overhead mixer at 23 °C for one hour. Following, the reaction was stopped by centrifugation (5 min, 1,600 g). Buffer solution (ph 7.8, mixed from Na_2HPO_4 83.6 mmol L^{-1} and KH_2PO_4 66.7 mmol L^{-1}) was pipetted into cuvettes and filled up and mixed with a part of the supernatant of the sample (FDA = 1.8 ml buffer and 0.2 ml sample; MUF-P = 0.25 ml buffer and 1 ml 1:20 pre-diluted sample). The fluorescence was detected with a fluorescence spectrometer at 470 nm (emission 513 nm) for FDA and at 365 nm (emission 450 nm) for MUF-P.

Statistical analyses were performed with the software SPSS (version 9.0.1, SPSS Inc., USA). Arithmetic means with standard deviations (mean ± SD) were calculated. In the figures, arithmetic means with minimum and maximum values (mean ± min/max) were presented.

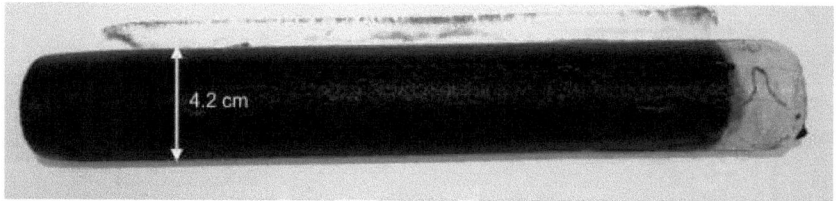

Fig. III-9. Frozen sediment core for the sampling of the microbial analyses.

Results

The total bacterial abundance estimated by the DAPI counting method (Fig. III-10) resulted in higher abundances in the sediment surface (n=2; 2.96 x 10^9) and the burrow walls (n=2; 2.39 x 10^9) than in the non-affected sediment (n=2; 6.31 x 10^8).

The Live/Dead *Back*Light Kit (Fig. III.11) resulted also in higher bacterial abundances in the sediment surface (n=2; live: 7.77 x 10^8, dead: 7.07 x 10^8) and the burrow walls (n=2; live: 7.66 x 10^8, dead: 6.85 x 10^8) than in the non-affected sediment (n=2; live: 5.93 x 10^8, dead: 3.85 x 10^8). This result was detected for both, living and dead bacteria. Besides, the relation between living and dead bacteria was higher for the non-affected sediment (1.54) than for the burrow walls (1.12) and the sediment surface (1.10).

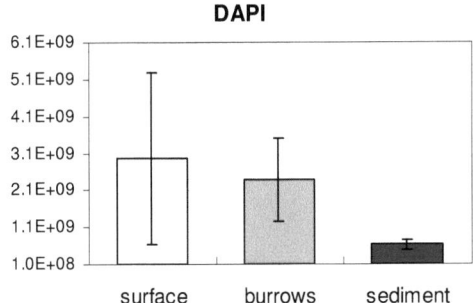

Fig. III-10. Total bacterial abundance (mean ± min/ max) estimated by the DAPI counting method.

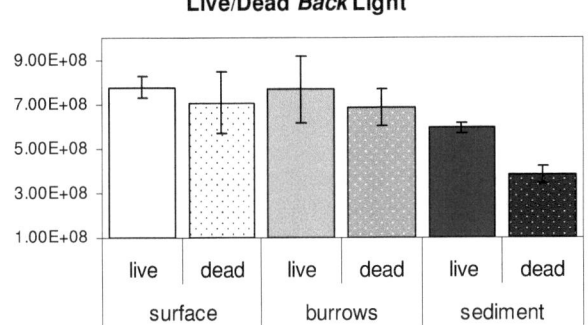

Fig. III-11. Bacterial abundance (mean ± min/ max) of live as well as of dead bacteria estimated by the Live/Dead BackLight Kit.

The identification of the microbial community structure via DGGE resulted in specific communities for all three sediment areas. The NMS-analysis (Fig. III-12 a and b) demonstrated that the cDNA (active bacteria) amplified from the burrow walls was more similar to the cDNA from the sediment surface than to the cDNA from the non-affected sediment (n=2 each). This result was detected for both, Eubacteria and Archaea.

The NMS-analysis (Fig. III-12 a and b) of the DNA (present bacteria) shows relatively variable results for the three sediment areas regarding to Eubacteria. In contrast, for Archaea the DNA-genes amplified from the burrow walls were more similar to the non-affected sediment than to the sediment surface (n=2 each).

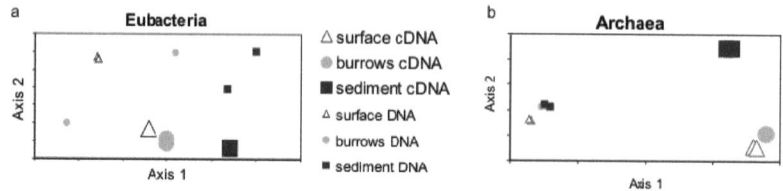

Fig. III-12. NMS-analysis (non metric scaling) for cDNA (active bacteria communities) and DNA (present bacteria communities) for a) Eubacteria and b) Archaea extracted for material from sediment surface, burrow walls and non-affected sediment (n=2 each).

The extra cellular enzymatic activities determined by the activities of general hydrolases (lipase, protease, and esterase) (Fig. III-12) resulted in significant higher values for the sediment surface (n=3; 1.94 ± 0.08 mmol FDA g^{-1} DW h^{-1}) (U-test, p=0.05) and the burrow walls (n=6; 1.8 ± 0.08 mmol FDA g^{-1} DW h^{-1}) (U-test, p=0.020) than for the non-affected sediment (n=3; 1.47 ± 0.02 mmol FDA g^{-1} DW h^{-1}).

Enzymatic activities of the phosphatases (Fig. III.13) were significantly higher (mean ± SD) in the sediment surface (n=3; 7.13 ± 0.1 mmol MUF-P g^{-1} DW h^{-1}) than in the non-affected sediment (n=3; 6.87 ± 0.09 mmol MUF-P g^{-1} DW h^{-1}) (U-test, p=0.050). The burrow walls (n=6; 6.93 ± 0.4 mmol MUF-P g^{-1} DW h^{-1}) do not differ significantly from the two other sediment areas.

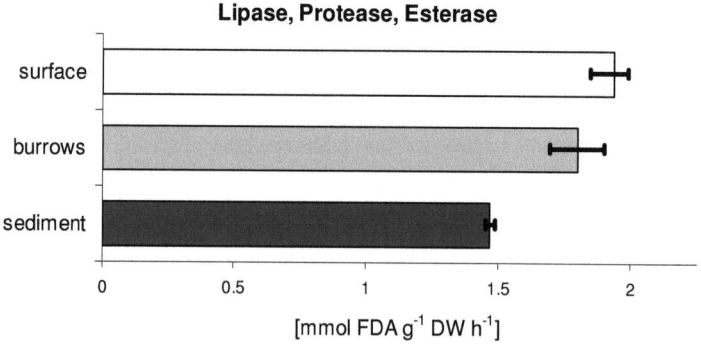

Fig. III-13. Extra cellular enzymatic activities determined by the activities of general hydrolases (lipase, protease, and esterase) (mean ± min/ max).

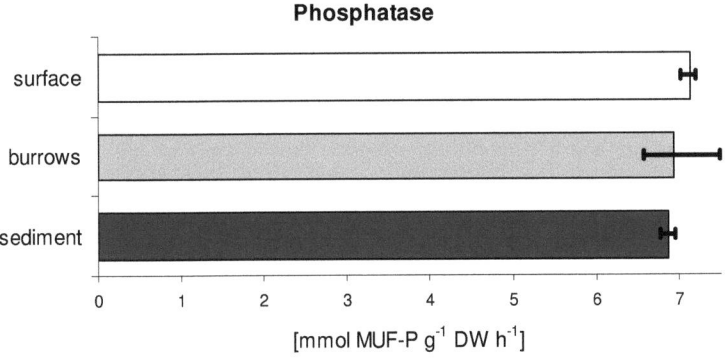

Fig. III-14. Extra cellular enzymatic activities determined by the activities of phosphatases (mean ± min/ max).

Discussion

The analysis of microbial abundance, community structure and activities clearly show that bioirrigating *C. plumosus* larvae affect the sediment microbiology.

In comparison to the non-affected anoxic sediment, the bacterial abundance in the oxidized burrows walls is enhanced. The bacterial numbers counted in the walls are similar to the oxic sediment surface. Besides, the relation of live and dead bacteria is much lower in the burrow walls than in the non-affected sediment and comparable to the sediment surface. Consequently, it is conjectural that bioirrigation and other activities performed by *C. plumosus* larvae stimulate the bacterial growth and the reproduction of certain bacterial species.

The identification of the microbial community structure resulted in a similar outcome since the communities of active Eubacteria and Archaea extracted from the burrow walls are much more similar to the oxic sediment surface than to the adjacent anoxic sediment. An identification of the present Eubacteria and Archaea (including DNA of active as well as inactive (dead) bacteria) indicates that the community structure of the oxidized burrow walls is recently shifted to the community structure of the oxic sediment surface. The present Archaea in the burrow walls are more similar to the communities in the non-affected sediment than to the sediment surface. The present Eubacteria in the burrow walls and in the non-affected sediment show a relatively high variability and are quite different from the sediment surface.

From literature it can be assumed that in freshwater sediments the microbial biomass as well as the community respiration are increased due to bioirrigation (Pelegri et al., 1994;

Stief et al., 2004). The authors suggest that the pore water flushing of amphipods or *C. riparius* larvae stimulate the growth and activity of sediment microorganisms such as nitrifiers and denitrifiers via an increased supply of O_2 and NO_3^-. Equally, bacterial cell counts in the burrow walls of the marine worms *N. virens* and *N. diversicolor* are higher than in the sediment surface and much more higher than in the non-affected sediment (Papaspyrou et al., 2006). However, the congruence of the community structures of burrow walls and sediment surface are lower than of burrow walls and ambient non-affected sediment. In contrast, the microbial communities in the burrows of a marine bioirrigating ghost shrimp are very similar to the surface communities and are very different from communities deeper in the sediment (Bertics & Ziebis, 2009). The microbial communities in the burrow walls of fiddler crabs, which are not actively ventilating their burrows, do not differ significantly from communities in the non-affected sediment collected from the same sediment depth (Bertics & Ziebis, 2009). For both crustaceans, the microbial abundance in the burrow walls increases in comparison to the sediment that is not affected by burrowing. However, these examples show that different burrowing invertebrate in different environments affect the microbial communities in different ways. Furthermore, the burrow walls have characteristics (alternating oxic-anoxic conditions, organic content, and sediment structure) that are not comparable to the reducing conditions of the adjacent sediment. Nevertheless, burrow walls should not be considered as a simple extension of the oxic sediment surface since the periodically irrigated milieu of the burrow linings is quite different from the overlying water body. However, oxygen can be identified as a key parameter structuring microbial communities because it is the most favorable electron acceptor for carbon oxidation in sediments. So far, many key parameters responsible for the microbial community structure and its functions are not clarified. Moreover, the effects of bioirrigation and bioturbation have to be regarded separately for every sediment, different species and diverse environmental conditions.

The analyses of the general hydrolases (lipase, protease, and esterase) and phosphatases indicate that the extra cellular enzymatic activities in the oxic sediment surface are considerably higher than in the anoxic non-affected sediment, while the oxidized burrow walls of *C. plumosus* are quite similar to the surface. For instance, Reichardt (1988) detected highest levels of certain hydrolytic enzymes (alkaline phosphatase and sulfatase) in the burrow walls of the marine lugworm *A. marina*. Stief (2007) has shown that in freshwater sediment various extra cellular enzymatic activities (α-glucosidase, β-glucosidase, and aminopeptidase) were significantly higher in the presence of *C. riparius* larvae than in their absence. However, the hydrolysis of organic macromolecules by microbial extra cellular enzymes is considered the rate-limiting step in mineralization of organic matter in aquatic environments (Hoppe, 1983; Meyer-Reil, 1987). Indeed, the interpretation of enzymatic activities has to be done carefully. First, the data presented in this study are not measured *in situ*. Besides, high concentrations of substrate are provided to assure no substrate limitation during the time of incubation. Thus, despite short incubation times, it cannot be excluded that such concentrations induce physiological changes in microorganisms (Stief, 2007). In contrast, enzyme inhibitors, the adsorption of substrate analogues, or the liberation of fluorophores to sediment particles lead to underestimated enzymatic activities (Coolen & Overmann, 2000).

In conclusion, microbial abundance and community structures are considerably affected by bioirrigation. Microbial analyses have shown that the oxidized burrow walls are more similar to the oxic sediment surface than to the non-affected anoxic sediment. Furthermore, enzymatic activities (general hydrolases, phosphatases) are higher in burrow walls than in non-affected sediment. However, as indicated by other authors, the influence of bioirrigation has to be regarded separately for the diverse aquatic environments. Besides, further investigations are necessary to identify the active microbial groups and its functions in the periodically oxic/ anoxic milieu of the burrow walls. As already discussed in chapter III.I, the ion concentrations in the pore water around the burrow linings are partially changed due to enhanced microbial processes. Hence, the effects of modified community structures and bioirrigation-mediated or rather stimulated microbial processes should have important effects on nutrient cycling in aquatic environments.

IV. Synopsis

IV.I. Summary of the results

As a result of this dissertation, several techniques appropriate to measure bioirrigation activity in U-shaped narrow burrows (diameter of ≤ 2 mm) in muddy sediment can be suggested. Successfully tested measurement techniques are video analysis, color tracers, O_2 microelectrodes, flow velocity microelectrodes, thermal flow sensors, PIV, and conductivity exchange experiments. In Table II-I-4, the suitability and simplicity of the techniques for the determination of the parameters flow velocity, pumping period frequency, pumping period duration, individual pumping time, and individual pumping rate are summarized. *(Chapters II.I and II.II)*

For subsequent investigations color tracers and PIV are chosen to quantify the flow velocity in a burrow, and O_2 and flow velocity microelectrodes are preferred to determine pumping period frequency and duration as well as the individual pumping time of *C. plumosus* larvae. Conductivity exchange experiments are used to quantify the pumping rate of the invertebrates directly. Additionally, the pumping rate is calculated from the results of the aforementioned measurements by multiplying the parameters individual pumping time and flow velocity with the cross-section of the burrows. *(Chapters II.I, II.II, II.III, and II.V)*

Reliable results measured with the aforementioned techniques (compare Table II-I-3) are taken for the calculation of average values (mean ± SD). Hence, for 4th instar *C. plumosus* larvae, a flow velocity of 14.9 ± 1.7 mm s^{-1} (color tracers, O_2 microelectrodes, flow velocity microelectrodes, PIV), a pumping period frequency of 26 ± 1 h^{-1} (flow velocity microelectrodes, thermal flow sensor), a pumping period duration of 1:32 ± 0:00 min:sec (flow velocity microelectrodes, thermal flow sensor), and an individual pumping time of 33 ± 3 min h^{-1} (O_2 and flow velocity microelectrodes, thermal flow sensor) is determined. An individual pumping rate of 61 ± 8 ml h^{-1} (color tracers, O_2 microelectrodes, flow velocity microelectrodes, thermal flow sensor, PIV, conductivity exchange) for a single larvae is measured when an average burrow diameter of 1.7 mm is assumed. This value results in a population pumping rate of 1.09 m^3 m^{-2} d^{-1} when 745 larvae per m^2 (see Appendix A) and a mean temperature of 20 °C is taken into account. *(Chapters II.I and II.III)*

For flow velocity measurements with color tracers, it is necessary to know the length of a macrozoobenthos burrow. By using setups of sediment-filled Perspex and wire tubes, the length of a burrow is predetermined. Consequently, measurements can be performed without an additional technique and the natural environment of the larvae is almost guaranteed. Setups of water-filled Tygon tubes are mainly used to observe and qualify the general behavior of an organism, as well as to assess the results of former studies using such an artificial setup. However, the flow velocity is decreased and the frequency and length of pumping are changed in artificial water-filled tubes. Thus, those setups are not appropriate for the quantitative determination of the aforementioned parameters. *(Chapters II.I and II.III)*

Length and shape of burrows directly built in the sediment are made visible by using X-ray analysis. To enhance contrast between sediment and burrow linings, Mo_2C that is mixed into the sediment is used as radio-opaque material. If the position of an organism in the sediment should be made visible as well, the body has to be marked with silver conductive paints before putting it into the mesocosm (see Appendix B). After the burrows are built, the mesocosm has to be carried carefully to the X-ray device for imaging. *(Chapter II.III)*

This study confirms that filter-feeding significantly increases the volume of water pumped by tube-dwelling macrozoobenthos. The pumping rate of *C. plumosus* larvae that perform pumping for the supply of oxygen and food is five times higher than the pumping rate of pupae that have to pump water only for respiration. *(Chapter II.II)*

Occasionally highly variable flow velocities are generated by *C. plumosus* larvae since also other activities such as feeding, moving, constructing, or cleaning induce a comparatively low or high water flow in the burrow. Such a flow may also occur in the opposite direction, from outlet to inlet. Moreover, the PIV measurements demonstrate the filtering effect of filter-feeders such as *C. plumosus* since the number of tracer particles decreased drastically during the experiments. *(Chapters II.II and II.III)*

To detect diffusive and advective transport processes through the burrow walls and into the adjacent sediment pore water, the three-dimensional nuclear medical imaging technique PET/CT (positron-emission-tomography/computer tomography) is applied successfully. [18F]fluoride is used as radioactive tracer since it is comparatively inert to the uptake by sediment microorganisms. The PET experiments clearly demonstrate that bioirrigation causes an advective transport into the sediment because at the side of the burrow outlet the tracer penetrates deeper into the sediment than at the side of the burrow inlet (see Appendix C). Pressure differences along the course of the burrow can be used to separate between diffusive and advective transport. *(Chapter II.IV)*

The bioirrigation-caused influx of water and dissolved substances from the overlying water body into the sediment is analyzed with tracer experiments by using NaCl as an inert tracer. The influx rate, which is calculated with a mass balance equation, demonstrates that rising temperatures (from 10 to 20 °C) result in a clearly enhanced influx into the sediment (Q_{10}=1.7). *(Chapter II.V)*

The study of the bioirrigation activity of *C. plumosus* via changing environmental parameters indicates that bioirrigation is highly susceptible. The pumping rate of the larvae increases with rising temperature ($Q_{10}=1.4$) and dropping oxygen concentrations of the overlying water. Furthermore, the rate of water pumped through the burrows and into the adjacent sediment varies seriously with season. Despite constant holding conditions in the laboratory varying influx rates are measured throughout several years. *(Chapter II.V)*

The analysis of the pore-water species SRP, NH_4-N, Fe^{2+}, and SO_4^{2-} with a spatial high resolved 2D peeper technique shows a very heterogenic ion distribution in the sediment. The concentrations of SRP, NH_4-N, and Fe^{2+} decrease, whereas the concentration of SO_4^{2-} increases along the burrow linings in comparison to the non-affected sediment. Advective transport due to pressure differences inside the burrow lumen and diffusive transport due to concentration gradients between burrow water and pore water redistribute the ions in the sediment. Therefore, burrow-flushing directly causes alterations with vicinity to the burrow walls. Moreover, burrow-flushing induces the heterogeneity of the sediment by an oxidation of the burrow walls. Consequently, the redox milieu is altered and biogeochemical processes are affected in bioirrigated sediments. *(Chapter III.I)*

The analysis of P in bioirrigated sediments shows considerably increased values in the sediment matrix of the burrow walls in comparison to the sediment which is not affected by bioirrigation. The P-fractionation of samples from the oxic sediment surface, the oxidized burrow walls, and the anoxic non-affected sediment shows that in the oxic areas P precipitates with oxidized Fe^{3+} or Mn^{3+}. In contrast, less S is precipitated in the oxic sediment areas than in the anoxic ones. However, a high amount of P is fixed in the burrow walls. It can be assumed that P is being transported into the burrow walls from the overlying water body due to pumping activity of the larvae, as well as from the adjacent sediment due to diffusive and advective processes. Hence, in this study the retention of P is enhanced in bioirrigated sediment. *(Chapter III.II)*

Bacterial numbers are higher in the burrow walls than in the adjacent non-affected sediment, but quite similar to the sediment surface. This result is shown for the total bacterial abundance as well as for living bacteria. The active microbial community structure (Eubacteria and Archaea) in the burrow walls is also similar to the sediment surface, whereas the present community is variable and more similar to the adjacent sediment. The extra cellular enzymatic activities are generally higher in the sediment surface and in the burrow walls than in the non-affected sediment. *(Chapter III.III)*

Transport processes and significant changes caused by the bioirrigation activity of *C. plumosus* larvae at the sediment-water interface are summarized in Fig. IV-1.

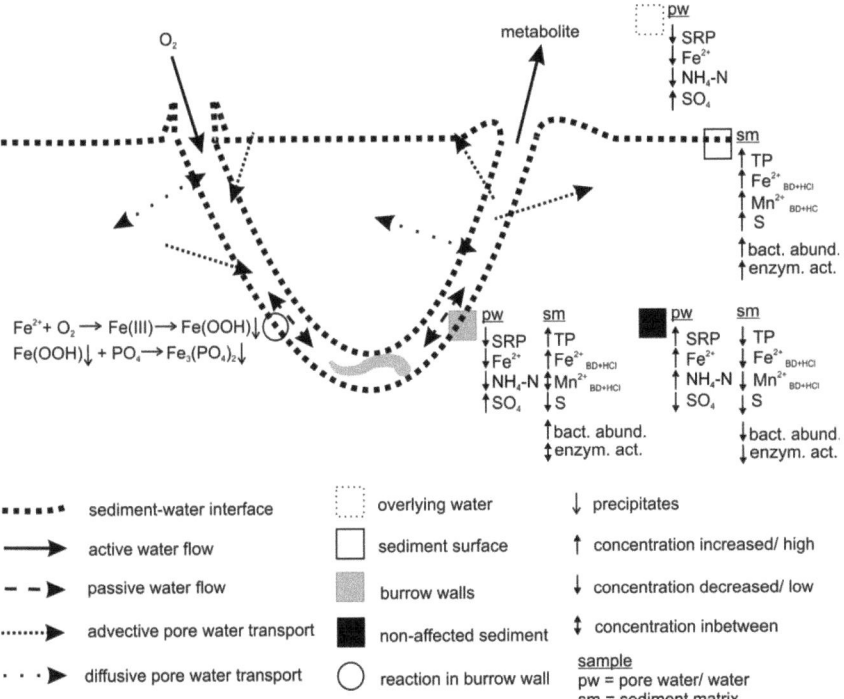

Fig. IV-1. Transport processes and changes in the sediment-water chemistry and microbiology (bacterial abundance and enzymatic activity) caused by a bioirrigating C. plumosus larva inside its burrow.

IV.II. Conclusions

The characterization of the hydrodynamic transport caused by macrozoobenthos dwelling in narrow burrows in muddy sediments and its biogeochemical impacts revealed a number of important findings. In the following paragraphs, these findings will be reviewed with respect to the four hypotheses postulated in the introduction.

Hypothesis 1 - Bioirrigating species dwelling in narrow burrows are able to cause relevant pumping rates: Former studies dealing with bioirrigation-caused fluxes at the sediment-water interface often focused on pumping rates of relatively big marine organisms such as *A. marina* or *U. caupo*. Chapters II.I to II.III of this study show that also the relatively small *C. plumosus* larvae dwelling in narrow burrows with a diameter of ≤ 2 mm create impressing pumping rates. A single 4th instar larva pumps a water volume of approximately 61 ml h^{-1} with an average flow velocity of 14.9 mm s^{-1} through its burrow. Nevertheless, the obviously bigger worms *A. marina*, *N. diversicolor*, and *U. caupo* are able to accomplish even higher pumping rates of approximately 200, 540, and 16,000 ml h^{-1} (Krüger, 1964; Riisgard, 1991; Osovitz & Julian, 2002). By comparison, larvae of *H. limbata* and *S. velata*, which are comparable in body size and burrow diameter to *C. plumosus* larvae, pump water with maximal 2 mm s^{-1} and 5.5 mm s^{-1}, respectively, and cause comparatively low pumping rates of less than 25 ml h^{-1} (Gallon et al., 2008).

However, individual pumping rates do not only depend on body size, but also on the feeding mode of the species. In contrast to *H. limbata*, *S. velata*, and *A. marina*, the macrozoobenthos species *C. plumosus*, *N. diversicolor*, and *U. caupo* are filter-feeders that have to pump water not only for respiration, but also for the supply of food. The clear increase of the pumping rate due to filter-feeding is confirmed in this study since the PIV measurements presented in chapter II.III demonstrate that the pumping rate of *C. plumosus* pupae is significantly lower than of larvae. A filter-feeding caused increase of the pumping rate is also proven in other studies (Riisgard & Larsen, 2005).

Besides, the pumping rate of a population is highly influenced by the density of the individuals. 745 *C. plumosus* larvae per m^2 determined in this study (see Appendix A) is a common population density (McLachlan, 1977; Gallepp, 1979; Tellioglu et al., 2008). During summer (at 20 °C), such a population is able to pump a water volume of 1.3 m^3 m^{-2} d^{-1}. Hence, the entire water volume of Lake Müggelsee (36.5 x 10^6 m^3) in Berlin, Germany, is theoretically pumped through the burrows of larvae living in the lake sediment within less than five days (compare chapter II.III). By comparison, a representative population of 2,400 *N. diversicolor* per m^2 is pumping 9.8 m^3 m^{-2} d^{-1} (Riisgard, 1991), whereas 61 *U. caupo* per m^2 cause a pumping rate of 23 m^3 m^{-2} d^{-1} (Osovitz & Julian, 2002). However, the pumping rate generated by *C. plumosus* larvae dwelling in narrow burrows is high in comparison to those much bigger species. Consequently, also small bioirrigating species are able to cause relevant pumping rates.

Hypothesis 2 - Advective pore water transport is not negligible in muddy sediments:
In literature the relevant transport processes in muddy and sandy sediments are controversially discussed. Some authors assume that burrow walls in muddy sediments are nearly impermeable because of high hydromechanical dampening (Aller, 1982; Meysman et al., 2006a). Moreover, some invertebrates such as *C. plumosus* larvae are coating the burrow walls with a mucous net to stabilize their burrows in the soft sediment (Leuchs & Neumann, 1990; Zorn et al., 2006). Due to the impermeable sediment structure

and the sealed burrow walls, the hydrodynamic exchange of sediment pore water and burrow lumens through the burrow walls would be lower in muddy than in sandy sediments. As a consequence, the sediment around the burrow linings would consume less oxygen. Hence, the bioirrigation-mediated biogeochemical effects would be very low in muddy sediments in comparison to sandy sediments. However, in literature commonly diffusion is assumed to be the only transport process in muddy sediments (Meysman et al., 2006a).

The PET measurements presented in chapter II.IV have clearly shown that there is a significant transfer of tracer from the burrow lumens, through the burrow walls and into the adjacent sediment, although the sediment of Lake Müggelsee is muddy (grain size < 63 μm (Kozerski & Kleeberg, 1998), k_f-value $4.85 \pm 1.19 \times 10^{-6}$) (see Appendix D). From the PET image it is obvious (see Appendix C) that in the part of the burrow outlet the penetration into the sediment is deeper and, therefore, faster than in the part of the burrow inlet. It can be assumed that this result is caused by pressure gradients inside the burrow depending on the position of the larvae. Compared with the adjacent sediment, pumping activity causes a lowered pressure in the inlet branch in front of the larva and an increased pressure in the outlet branch behind the larva. Consequently, in the outlet part of a burrow advective transport is directed from the burrow into the sediment and, therefore, added to diffusion. In the inlet part of a burrow, advective transport occurs from the adjacent sediment into the burrow and due to this counteracts diffusion. Nevertheless, it should be respected that the pumping characteristics of different invertebrates as well as the shape of their burrow system creates different pressure gradients inside the burrows (Forster-Smith, 1978), what may affect the surrounding sediment varyingly strong.

The results of the PET measurements are confirmed with the tracer experiments presented in chapter II.V. In this experiments, the influx rate increases significantly with increasing flow velocity, pumping rate, and density of *C. plumosus* larvae. Although diffusion increases with rising temperatures and strengthened concentration gradients, the impact of an intensified pumping is expected to be lower when diffusion would be the only transport process. To conclude, in this study it is proven that, depending on pump characteristics and sediment porosity, advection occurs in muddy sediments and should not be neglected.

Hypothesis 3 - Bioirrigation activity is highly sensitive to environmental changes:
So far, the effects of environmental changes on bioirrigation activity, pumping rate, and water influx into the sediment are rarely investigated. Most of the available studies

determine pumping rates depending on a single temperature and at a sufficient oxygen concentration. Studies examining effects of season are even scarcer (Schlüter et al., 2000).

Chapter II.V of this study shows that bioirrigation activity is highly sensitive to environmental changes such as rising temperature, dropping oxygen concentration, and season of the year. With dropping oxygen concentrations of the overlying water, the flow velocity generated by *C. plumosus* larvae generally decreases and the rhythm of pumping is being altered. The individual pumping time of the larvae increases down to an oxygen saturation of 3 %, below 2 % the pumping activity is nearly abandoned.

Rising temperatures (from 10 to 20 °C) significantly increase the flow velocity generated by *C. plumosus* (Q_{10}=1.9), whereas the time the larvae pump per hour remains constant. However, the rhythm of pumping is significantly modified with rising temperature since the length of single pumping periods decreases while the frequency of pumping significantly increases. The increasing flow velocity at warmer temperatures is responsible for a significant increase of pumping rates as well as influx rates of water and solutes into the sediment.

Bioirrigation activity is also sensitive to seasonal trends. Influx rates of overlying water into the sediment throughout the year are presented in chapter II.V (see also Appendix C). The investigation shows that the influx rates measured over almost three years are always higher in spring and early summer than in autumn. Under constant laboratory conditions, the same significant seasonal trend is observed for measurements at 10 and 20 °C. Consequently, despite ambient conditions such as temperature, oxygen content, and light climate a circannual cycle of the bioirrigation activity of *C. plumosus* larvae is shown in this study. In chapter II.V it is supposed that an interaction of endogenous clocks and external stimulations (Saunders et al., 2004) are responsible for the seasonal variations. Furthermore, it is being suggested that the influx rate of overlying water into the sediment is not only conditional on pumping rates, but also depending on other factors such as population density, food sources, sediment porosity, and burrow shape and dimensions. To summarize, this study shows that bioirrigation activity is highly sensitive to environmental changes. However, although water fluxes caused by bioirrigation activity are variable throughout the year, the exchange at the sediment-water interface is considerably enhanced in bioirrigated sediments.

Hypothesis 4 - Bioirrigation has a significant impact on the sediment biogeochemistry and leads to enhanced P retention in the sediment: The impact of tube-dwelling macrozoobenthos on the sediment chemistry is investigated in a number of previous studies (Matisoff et al., 1985; Andersen & Kristensen, 1991; Volkenborn et al., 2007). Nevertheless, analyses of the heterogeneity of bioirrigated sediment with a high spatial resolution are comparatively rare in literature.

Chapter III.I of this study confirms the findings of Lewandowski et al. (2005a; 2005b; 2007) that the concentrations of the pore water species SRP, NH_4-N, and Fe^{2+} increase with distance to the burrow walls. The ions are being transported out of the sediment and into the burrow lumens by diffusive and advective processes and out of the burrows by active pumping of the bioirrigating species. Diffusion is caused by concentration gradients between the sediment pore water (high concentration) and the water in the burrow that originates from the overlying water body (low concentration). Advection is caused by pressure gradients inside the burrow due to pumping activity (compare chapter II.III). Besides, the flushing of the burrow lumens with water from the overlying water body oxidizes the burrow walls. As a consequence, redox changes are responsible for a decrease of the concentrations of pore water ions. By contrast, concentrations of SO_4^{2-} are higher in the overlying water than in the sediment pore water as shown in chapter III.I. Thus, SO_4^{2-} is transported from the overlying water body into the burrow lumen by active pumping and penetrates through the burrow walls into the adjacent sediment by diffusive and advective processes. Furthermore, the SO_4^{2-} in the burrow walls originates from re-oxidation of S^{2-} that is transported by diffusion from the adjacent sediment pore water.

In literature it is suggested that the availability of oxidants (oxygen, nitrate, ferric iron) plays a key role in determining the presence and abundance of different microbial taxa (Bertics & Ziebis, 2009). In chapter III.III it is being demonstrated that the microbial analysis of burrow walls oxidized by *C. plumosus* larvae differs from the non-affected sediment. The bacterial abundance in the burrow walls is higher than in the adjacent anoxic sediment. Moreover, the microbial community structure of the burrows is more similar to the oxic sediment surface. The enzymatic activities are also affected by bioirrigation since the hydrolytic potential, which is considered as the rate-limiting step in organic matter mineralization (Hoppe, 1983; Meyer-Reil, 1987), is higher in the burrow walls than in the adjacent sediment. Thus, it can be assumed that the stimulated and enhanced microbiological processes in bioirrigated sediments affect the concentrations of pore water ions and further biogeochemical processes.

Both, the release and the retention of P due to bioirrigation activity are being demonstrated in literature. In a number of studies bioirrigation enhances the flux of P from the sediments into the overlying water body due to a mix of burrowing, bioirrigation, and excretion of the invertebrates (Gallepp, 1979; Graneli, 1979b; Mermillod-Blondin et al., 2005; Caliman et al., 2007; Chaffin & Kane, 2010). In studies with *Chironomus* species, the rates of P release depend on temperature, aeration of the sediment, density of the larvae, and chemical composition of pore water and overlying water (Gallepp, 1979; Callisto et al., 2009). The results presented in chapter III.II show retention of P due to bioirrigation activity. In the burrow walls of *C. plumosus* larvae significantly higher P concentrations are detected than in the non-affected sediment. In another experimental study (Ebeling, 2008) it is also shown that the P-absorption capacity of a sediment is clearly enhanced when *C. plumosus* is dwelling inside. Nevertheless, after hatching of the larvae, the burrows are filled up with sediment and become anoxic again. Subsequently oxidized iron compounds will be dissolved due to anoxic sediment conditions. Although the discoloration of the burrow walls indicating the oxygenation of the sediment is visible for many months, it can be assumed that subsequently the P diffuses from the sediment (high concentrations) to the overlying water body (low concentrations) by concentration gradients. However, for each sediment and bioirrigating species the potential for the release or the retention of P has to be evaluated separately. The sediments of Lake Müggelsee are rich in Fe^{2+} and P, and thus high capacities for the precipitation as $Fe(OOH)-PO_4$ are given. Besides, *C. plumosus* performs comparatively little bioturbation, but extensive bioirrigation, and their burrows reach deep into the anoxic sediment (20-30 cm). The relatively far distance to the sediment surface should reduce diffusive fluxes of P into the overlying water body after hatching of the larvae.

To sum up, the results of this dissertation show that bioirrigation might have a significant and far-reaching effect on the sediment biogeochemistry.

IV.III. Outlook

With the techniques described and used in chapter II useful tools for the qualification and the quantification of bioirrigation-caused hydrodynamic transport processes are provided. Moreover, the biogeochemical impacts of tube-dwelling *C. plumosus* larvae are shown in chapter III. In further studies, the developed techniques should be used to investigate the bioirrigation rates of other, especially small macrozoobenthos species. Additionally, a number of open questions with regard to bioirrigation can be answered with the techniques provided in this dissertation.

One open question is the impact of tube-dwelling macrozoobenthos on the filtering of the overlying water body. In case of high larvae densities, filter-feeding performed by *C. plumosus* might generate a high pressure on phyto- and zooplankton in shallow polymictic lakes. It is possible that the pumping drastically alters the plankton succession in polymictic lakes and effects the water quality as it is known for bivalves (Fanslow et al., 1995; Pusch et al., 2001). On the one hand, since phytoplankton is filtered from the water the larvae might delay or weaken phytoplankton blooms in eutrophic lakes. On the other hand, due to filtering zooplankton from lake water the larvae might decrease grazing pressure on phytoplankton and favor phytoplankton blooms. Further, in biofiltration systems suspension feeding organisms such as bivalves are used for the biological removal of particulate and dissolved nutrients from water (Shpigel, 2005). Perhaps it might be possible to use filter-feeders such as *Chironomus* larvae as biofilters, too. However, the effect of filter-feeding of tube-dwelling species can be investigated for instance with PIV or turbidity measurements.

Besides the effect of filtering particles out of the water, bioirrigation might enhance the resuspension of particulate matter from the sediment to the overlying water body. For marine species, the resuspension of particles is demonstrated in a few studies (Graf & Rosenberg, 1997; Volkenborn et al., 2007). Information about the context of resuspension and bioirrigation activity of *Chironomus* or other limnic species is scarce in literature. PIV measurements might be helpful to investigate the impact of bioirrigation on particle resuspension.

Recent studies (Stief et al., 2009; Stief & Schramm, 2010) demonstrate that the influence of bioirrigation on the cycling of N is often underestimated in limnic systems. Macrozoobenthos species emit N_2 and the greenhouse gas N_2O since it is produced in their gut and in the burrow walls by nitrifying and denitrifying bacteria. In bioirrigated sediments, denitrification is stimulated, and a periodic irrigation of the burrow causes fluctuating chemical conditions that are possible to increase the emission rate of N_2O (Schreiber et al., 2009). Thus, for *C. plumosus* the production of N_2O by sediment microorganisms is of greater importance than the emission by their gut. Here, further studies should tie up and investigate the impact of different bioirrigating species and sediments on the cycling of N. For the investigations, peeper analysis and the determination of pumping rates with the techniques described in chapter II.I should be useful.

Furthermore, the impact of bioirrigation on the microbial production of toxic substances and its release to the overlying water body is rarely investigated. Some studies show that in marine sediments the transformation and release of toxic substances such as monomethylmercury (MMHg) (pers. com. G. Nogaro, Department of Earth & Environmental Sciences, Wright State University, USA) or Tributyltin (TBT) (Hamer &

Karius, 2005) increase when macrozoobenthos is dwelling in the sediment. Information about the context of the release of toxic substances and the bioirrigation activity of limnic species in freshwater sediments are scarce in literature. Peeper analysis might be a helpful tool to investigate the presence of toxic substances in bioirrigated sediments.

The impact of *C. plumosus* larvae on the retention of P is confirmed in chapter III.II. It is not yet clarified how long the fixed P remains in the burrow walls when the sediment is reduced after disappearance of the larvae. Furthermore, it is unknown how fast the dissoluted P diffuses through the sediment and how long it takes until the P is completely released back into the overlying water body. It might also be possible that the P is embedded in the sediment due to changes of the binding form and the formation of minerals (Jones & Browser, 1978; Goldhammer et al., 2010). A long-term effect of bioirrigation on P retention could be effect the cycling of P in aquatic environments seriously. Thus, detailed information about the P in the sediment after the hatching of the larvae would be desirable. For the analysis of the P fluxes in the sediment, tracer experiments with PO_4 performed with PET or influx experiments may be applicable. Moreover, a fractionation of P should be performed in different intervals after the hatching of the larvae.

In order to understand the biogeochemical effects of bioirrigation, the way and frequency of the pumping activity of different bioirrigating species is highly important. A filter-feeding *C. plumosus* larva performs periodically pumping, and during pumping periods, the water is being pumped through the burrows with a relatively high flow velocity. Therefore, the gradients between the water in the burrow and the sediment pore water are high and diffusive exchange as well as biogeochemical reactions are accelerated. During non-pumping periods, the water remains in the burrow lumen for a considerably longer time than during pumping. Thus, reactions and diffusive exchange decline throughout a non-pumping period. The quality of such water must be markedly changed in comparison to the water coming out of the burrow during pumping. However, the composition of the out-coming water in order to the bioirrigation characteristics of different invertebrates is not investigated up to now. For the investigations, measurements with micro sensors might be appropriate.

As shown in this dissertation (chapter II.V) and other studies, environmental changes such as shifts in temperature, oxygen concentration, season, food availability, population density, or the presence of predators considerably influence the bioirrigation activity of tube-dwelling macrozoobenthos (Walshe, 1948; McLachlan, 1977; Leuchs, 1986; Schlüter et al., 2000; Stief & Hölker, 2006). Therefore, those parameters should be addressed more detailed when pumping rates of different species are determined. Besides, little is known about the endogenous clocks of the invertebrates and external stimulations that are controlling their pumping rates. Further laboratory studies can be performed with the measurement techniques presented in chapter II. *In situ* changes might be determined by analyzing the glycogen content of an organism. Glycogen is a key factor for the identification of the energy budget and vitality of an organism (Hamburger et al., 1995).

Although oxygen can be identified as an important key parameter structuring microbial communities and stimulating microbial turnover rates, further studies are necessary to identify the other key parameters which are also essential. With detailed analyses the active microbial groups in the sediment especially in the modified milieu of the burrow walls have to be identified. Besides, the functions of the active microorganisms in the periodically oxic/ anoxic burrow walls, and thus the effects of the stimulated processes on organic matter mineralization and nutrient cycling are scarcely investigated. Constitutive to the investigations performed in chapter III.III, clone libraries and phylogenetic trees should help to identify the diverse bacterial groups and their functions (Lehours et al., 2007). Additionally to the general hydrolases and phosphatases, the determination of further extra cellular enzymatic activities such as sulphatase, α-glucosidase, β-glucosidase, and aminopeptidase should help to identify modified microbial processes in the burrow walls (Reichardt, 1988; Marxsen et al., 1998; Stief & de Beer, 2006). The measurement of the intra cellular enzymatic activity via dehydrogenase activity or aerobic respiration and denitrification could help to assess the activity of the sediment microorganisms (Marxsen et al., 1998; Nogaro et al., 2007). Besides, the microbial production can be identified by the measurement of the microbial formation of protein via leucine incorporation (Marxen, 1996). The degradation of organic matter might be determined by iron, manganese, and sulphate reduction (Canfield et al., 1993).

The clarification of the role of bioirrigation for an aquatic environment might be important for recommendations for the restoration of eutrophicated lakes. On the one hand, in case of an aeration of the deep lake water, bioirrigating species might be stimulated. On the other hand, with measures such as the covering of the sediment surface with plastic foil or mineral substances or the addition of chemical substances, bioirrigating species might be eliminated. Furthermore, the restoration might be ineffective when high densities of bioirrigating and bioturbating species living inside. However, since bioirrigation also stimulates or eliminates the effects of eutrophication their role should be respected when a restoration is planned and implemented.

So far, with the hydrodynamic and biogeochemical data provided in chapter II and III, a limnic small-scale reactive transport model can be developed. This model should consider the effects of bioirrigation on biogeochemical reaction rates and microbial turnover, as well as on the exchange processes at the sediment-water interface. As shown in literature (Meile et al., 2003), modeling is highly important to understand transport processes of different pore water species to estimate the intensity of bioirrigation. Therefore, future research should focus on the interoperation of the measurement techniques presented above, with reactive transport modeling. Thus, new possibilities to understand the processes, relationships, and interactions in sediments can be provided and the role of bioirrigation for nutrient cycling and diagenesis can be investigated in detail for different aquatic environments.

References

Aller, 1978: Aller, R. C., 1978. The effects of animal-sediment interactions on geochemical processes near the sediment-water interface. In Wiley, M. L. (ed), Estuarine Interactions. Academic Press, New York: 157-172.

Aller, 1980: Aller, R. C., 1980. Quantifying solute distributions in the bioturbated zone of ma-rine sediments by defining an average microenvironment. Geochimica et Cos-mochimica Acta 44: 1955-65.

Aller, 1982: Aller, R. C., 1982. The effects of macrobenthos on chemical properties of marine sediment and overlying water. In McCall, P. L. & M. J. S. Tevesz (eds), Animal-Sediment Relations. Plenum Press, New York: 53-102.

Aller, 1983: Aller, R. C., 1983. The importance of the diffusive permeability of animal burrow linings in determining marine sediment chemistry, Journal of Marine Research 41: 299-322.

Aller, 1988: Aller, R. C., 1988. Benthic fauna and biogeochemical processes in marine sediments. In Blackburn, T. H. & J. Sorenson (eds). Nitrogen cycling in coastal marine environments. John Wiley and Sons, New York: 301-338.

Aller, 1994: Aller, R. C., 1994. Bioturbation and remineralization of sedimentary organic matter: effects of redox oscillation, Chemical Geology 114: 331-345.

Aller, 2001: Aller, R. C., 2001. Transport and reactions in the bioirrigated zone. In Boudreau, B. P. & B. B. Jorgensen (eds), The Benthic Boundary Layer. Oxford Univ. Press, New York: 269-301.

Aller & Aller, 1992: Aller, R. C.& J. Y. Aller, 1992. Meiofauna and solute transport in marine muds, Limnology and Oceanography 37: 1018-1033.

Aller & Aller, 1998: Aller, R. C. & J. Y. Aller, 1998. The effect of biogenic irrigation intensity and solute exchange on diagenetic reaction rates in marine sediments, Journal of Marine Research 56: 905-936.

Aller & Yingst, 1978: Aller, R. C. & J. Y. Yingst, 1978. Biogeochemistry of tube-dwellings: a study of the sedentary polychaete Amphitrite ornata (Leidy), Journal of Marine Research 36: 201-254.

Andersson et al., 1988: Andersson, G., W. Graneli, & J. Stenson, 1988. The Influence of animals on phosphorus cycling in lake ecosystems, Hydrobiologia 170: 267-284.

Andersen & Jensen, 1991: Andersen, F. O.& H. S. Jensen, 1991. The influence of chironomids on decomposition of organic matter and nutrient exchange in a lake sediment, Verhandlungen des Internationalen Verein Limnologie 24: 3051-3055.

Andersen et al., 2006: Andersen, F. O., M. Jorgensen & H. S. Jensen, 2006. The influence of *Chironomus plumosus* larvae on nutrient fluxes and phosphorus fractions in aluminium treated lake sediment, Water, Air, and Soil Pollution 6: 465-474.

Andersen & Kristensen, 1991: Andersen, F. O.& E. Kristensen, 1991. Effects of burrowing macrofauna on organic matter decomposition in coastal marine sediments, Symposium of the Zoological Society of London 63: 69-88.

Andersen & Kristensen, 2002: Andersen, M.& E. Kristensen, 2002. The importance of bacteria and microalgae in the diet of the deposit-feeding polychaete *Arenicola marina*, Ophelia 56: 179-196.

Anderson & Meadows, 1987: Anderson J. G, & P. S. Meadows, 1978. Microenvironments in marine sediments. Proceedings of the Royal Society of Edinburgh 76(B):1-16

Armitage et al., 1995: Armitage, P., P. S. Cransotn, & L. C. V. Pinder, 1995. The Chironomidae: biology and ecology of non-biting midges. Chapman and Hall, London, UK.

Augenfeld & Neess, 1961: Augenfeld, J. M. & J. C. Neess. 1961. Observations on the respiratory enzymes of various life-sages of *Chironomus plumosus*, *Chironomus stageri*, and *Aedes aegypti*. The Biological Bulletin 120:139.

Battin et al., 2008: Battin, T. J., L. A. Kaplan, S. Findlay, C. S. Hopkinson, E. Marti, A. I. Packman, J. D. Newbold & F. Sabater, 2008. Biophysical controls on organic carbon fluxes in fluvial networks, Nature Geoscience 1: 95-100.

Benoit et al., 1991: Benoit, J. M., T. Torgersen & J. O´Donnell, 1991. An advection/diffusion model for 222Rn transport in near-shore sediments inhabited by sedentary polychaetes, Earth and Planetary Science Letters 105: 463-473.

Berg & McGlathery, 2001: Berg, P.& K. J. McGlathery, 2001. A high-resolution pore water sampler for sandy sediments, Limnology and Oceanography 46: 203-210.

Bertics & Ziebis, 2009: Bertics, V. J.& W. Ziebis, 2009. Biodiversity of benthic microbial communities in bioturbated coastal sediments is controlled by geochemical microniches, The ISME Journal 3: 1269–1285.

Biswas et al., 2009: Biswas, J. K., S. Rana, J. N. Bhakta & H. S. Jensen, 2009. Bioturbation potential of chironomid larvae for the sediment-water phosphorus exchange in simulated pond systems of varied nutrient enrichment, Ecological Engineering 23: 1444-1453.

Bostrom et al. 1988: Bostrom, B., J. M. Andersen, S. Fleischer & M. Jansson, 1988. Exchange of phosphorus across the sediment-water interface, Hydrobiologia 170: 229-244.

Boudreau & Marinelli, 1994: Boudreau, B. P.& R. L. Marinelli, 1994. A modelling study of discontinuous biological irrigation, Journal of Marine Research 52: 947-968.

Boulos et al., 1999: Boulos, L., M. Prevost, B. Barbeau, J. Coallier & R. Desjardins, 1999. LIVE/DEAD BackLight: application of a new rapid staining method for direct enumeration of viable and total bacteria in drinking water, Journal of Microbiological Methods 37: 77-86.

Brand et al. 2007: Brand, A., B. Müller, A. Wüest & C. Dinkel, 2007. Microsensor for in-situ flow measurements in benthic boundary layers at sub-millimeter resolution with extremely slow flow, Limnology and Oceanography: Methods 5: 185-191.

Brennan & McLachlan, 1979: Brennan, A. & A. J. McLachlan, 1979. Tubes and tube-building in a lotic chironomid (Diptera) community, Hydrobiologia 67: 173–178.

Burger & Buck, 1997: Burger, C. & A. Buck, 1997. Requirements and implementation of a flexible kinetic modeling tool, Journal of Nuclear Medicine 38: 1818-1823.

Caliman et al. 2007: Caliman, A., J. J. F. Leal, F. A. Esteves, L. S. Carneiro, R. L. Bozelli & Farjalla.V.F., 2007. Functional bioturbator diversity enhances benthic-pelagic processes and properties in experimental microcosms, Journal of the North American Benthological Society 26: 450-459.

Callisto et al. 2009: Callisto, M., J. J. F. Leal, M. P. Figueiredo-Barros & F. d. Assis Esteves, 2009. Effect of bioturbation by *Chironomus* on nutrient fluxes in an urban eutrophic reservoir, Conference paper, 7th ISE & 8th HIC, Chile.

Canfield et al. 1993: Canfield, D. E., B. Thamdrup & J. W. Hansen, 1993. The anaerobic degradation of organic-matter in Danish coastal sediments: Iron reduction, manganese reduction, and sulfate reduction, Geochimica et Cosmochimica Acta 57: 3867-3883.

Chaffin et al. 2010: Chaffin, J. D.& D. D. Kane, 2010. Burrowing mayfly (*Ephemeroptera: Ephemeridae*: Hexagenia spp.) bioturbation and bioirrigation: A source of internal phosphorus loading in Lake Erie, Journal of Great Lakes Research 36: 57-63.

Charbonneau et al., 1997: Charbonneau, P., L. Hare & R. Carignan, 1997. Use of X-ray images and a contrasting agent to study the behavior of animals in soft sediments, Limnology and Oceanography 42: 1823–1828.

Charles et al., 2004: Charles, S., M. Ferreol, A. Chaumot & A. R. R. Péry, 2004. Food availability effect on population dynamics of the midge *Chironomus riparius*: a Leslie modelling approach, Ecological Modelling 175: 217–229.

Christensen et al., 2000: Christensen, B., A. Vedel & E. Kristensen, 2000. Carbon and nitrogen fluxes in sediment inhabited by suspension-feeding (*Nereis diversicolor*) and non-suspension-feeding (*N. virens*) polychaetes, Marine Ecology Progress Series 192: 203-217.

Coolen & Overmann, 2000: Coolen, M. J. L. & J. Overmann, 2000. Functional exoenzymes as indicators of metabolically active bacteria in 124,000-year-old sapropel layers of the eastern Mediterranean Sea, Applied and Environmental Microbiology 66: 2589-2598.

Davis et al., 1975: Davis, R. B., D. L. Thurlow & F. E. Brewster, 1975. Effects of burrowing tubiticid worms on the exchange of phosphorus between lake sediment and overlying water, Verhandlungen Internationale Vereinigung für Theoretische und Angewandte Limnologie 19: 382-394.

D´Andrea & Lopez, 1997: D´Andrea, A. F.& G. R. Lopez, 1997. Benthic macrofauna in a shallow water carbonate sediment: Major bioturbators at the Dry Tortugas, Geo-Marine Letters 17: 282.

Drieser et al., 1993: Driescher E, H. Behrendt, G. Schellenberger & R. Stellmacher, 1993. Lake Müggelsee and its environment – natural conditions and anthropogenic impacts, Internationale Revue der gesamten Hydrobiologie und Hydrographie 78: 327-343.

Ebeling, 2008: Ebeling, C., 2008. Einfluss von *Chironomus plumosus* auf die Phosphor-Festlegung in Sedimenten und auf Stofftransportprozesse zwischen Sediment und Wasser. Institut für Gewässerökologie und Binnenfischerei & Humboldt-Universität, Diplomarbeit, Berlin.

Eckert & Walz, 1999: Eckert B. & N. Walz, 1999. Zooplankton succession and thermal stratification in the polymictic shallow Müggelsee (Berlin, Germany): a case for the intermediate disturbance hypothesis? Hydrobiologia *387*: 199-206.

Edmondson & Lehman, 1981: Edmondson, W. T.& J. T. Lehman, 1981. The effect of changes in the nutrient income on the condition of Lake Washington, Limnology and Oceanography *26*: 1-29.

Edwards & Rolley, 1965: Edwards, R. W.& H. L. J. Rolley, 1965. Oxygen consumption of river mud, Journal of Ecology *53*: 1-19.

Edwards et al., 2009: Edwards, W. J., F. M. Sost, G. Marisoff & D. W. Schloesser, 2009. The effect of mayfly (*Hexagenia* spp.) burrowing activity on sediment oxygen demand in western Lake Erie, Journal of Great Lakes Research *35*: 507-516.

Epler, 2001: Epler, J.H., 2001. Identification manual for the larval Chironomidae (Diptera) of North and South Carolina. A guide to the taxonomy of the midges of the Southeastern United States, including Florida. Special Publication SJ2001-SP13. North Carolina Department of Environment and Natural Resources, Raleigh, N.C., and St. Johns River Water Management District, Palatka. Version 1.0, Genus Chironomus: 8.39-8.44.

DIN, 1999: Fachgruppe Wasserchemie in der Gesellschaft Deutscher Chemiker in Gemeinschaft mit dem Normenausschuß Wasserwesen (NAW) im Deutschen Institut für Normung e.V., 1999. VCH Verlagsgesellschaft mbH Deutsche Einheitsverfahren zur Wasser-, Abwasser- und Schlammuntersuchung, Beuth Verlag GmbH.

Famme et al., 1986: Famme, P., H. U. Riisgård & C. B. Jørgensen, 1986. On direct measurement of pumping rates in the mussel *Mytilus edulis*, Marine Biology *92*: 323-327.

Fanslow et al., 1995: Fanslow, D. L., T. F. Nalepa & G. A. Lang, 1995. Filtration rates of the zebra mussel (*Dreissena polymorpha*) on natural seston from Saginaw Bay, Lake Huron, Journal of Great Lakes Research *21*: 489-500.

Faul et al., 2009: Faul, F., E. Erdfelder, A. Buchner & A.-G. Lang, 2009. Statistical power analyses using G*Power 3.1: tests for correlation and regression analyses, Behavior Research Methods *41*: 1149-1160.

Fay, 1994: Fay, J. A., 1994. Introduction to fluid mechanics, MIT Press, Cambridge, Massachusetts.

Fischer, 1982: Fischer, J. B., 1982. Effects of macrobenthos on the chemical diagenesis of freshwater sediments. In MCall, P. L. & M. J. Tevesz (eds), The biotic alteration of sediments, Plenum. New York: 177-218.

Forster & Graf, 1995: Forster, S.& G. Graf, 1995. Impact of irrigation on oxygen flux into the sediment: intermittent pumping by *Callianassa subterranea* and "piston-pumping" by *Lanice conchilega*, Marine Biology *123*: 335-346.

Forster-Smith, 1978: Forster-Smith, R. L., 1978. An analysis of water flow in tube-living animals, Journal of Experimental Biology and Ecology *34*: 73-95.

Forster-Smith & Shillaker, 1977: Foster-Smith, R. L.& R. O. Shillaker, 1977. Tube irrigation by *Lembos websteri* Bate and *Corophium bonnelli* Milnne Edwards (Crustacea: Amphipoda), Journal of Experimental Marine Biology and Ecology *26*: 289-296.

Francois et al., 2002: Francois, F., M. Gerino, G. Stora, J. P. Durbec & J.-C. Poggiale, 2002. Functional approach to sediment reworking by gallery-forming macrobenthic organisms: modeling and application with the polychaete *Nereis diversicolor*, Marine Ecology Progress Series *229*: 127-136.

Furukawa et al., 2001: Furukawa Y., S. J. Bentley & D. L. Lavoie, 2001. Bioirrigation modelling in experimental benthic mesocosms, Journal of Marine Research *59*: 417-452

Gallepp, 1979: Gallepp, G. W., 1979. Chironomid influence on phosphorus release in sediment-water microcosms, Ecology *60*: 547-556.

Gallepp et al., 1978: Gallepp, G. W., J. F. Kitchell & S. M. Bartell, 1978. Phosphorus release from lake sediments as affected by chironomids, Verhandlungen Internationale Vereinigung für Theoretische und Angewandte Limnologie *20*: 458-465.

Gallon et al., 2008: Gallon, C., L. Hare & A. Tessier, 2008. Surviving in anoxic surroundings: how burrowing aquatic insects create on oxic microhabitat, Journal of the North American Benthological Society *27*: 570-580.

Gerten & Adrian, 2002: Gerten, D. & R. Adrian, 2002. Effects of climate warming, North Atlantic Oscillation and El Niño on thermal conditions and plankton dynamics in European and North American lakes, The Scientific World Journal *2*: 586-606.

Gilbert et al., 1998: Gilbert, F., G. Stora & P. Bonin, 1998. Influence of bioturbation on denitrification activity in Mediterranean coastal sediments: an in situ experimental approach, Marine Ecology Progress Series *163*: 99-107.

Goldhammer et al., 2010: Goldhammer, T., V. Brüchert, T. G. Ferdelman & M. Zabel, 2010. Microbial sequestration of phosphorus in anoxic upwelling sediments, Nature Geoscience *3*: 557-561.

Graf & Rosenberg, 1997: Graf, G.& R. Rosenberg, 1997. Bioresuspension and biodeposition: a review, Journal of Marine Systems *11*: 269-278.

Graneli, 1979a: Graneli, W., 1979a. The influence of *Chironomus plumosus* larvae on the oxygen-uptake of sediment, Archiv für Hydrobiologie *87*: 385-403.

Graneli 1979b: Graneli, W., 1979b. The influence of *Chironumus plumosus* larvae on the exchange of dissolved substances between sediment and water, Hydrobiologia *66*: 149-159.

Grey et al., 2004: Grey, J., A. Kelly, S. Ward, N. Sommerwerk & R. I. Jones, 2004. Seasonal changes in the stable isotope values of lake-dwelling chironomid larvae in relation to feeding and life cycle variability. Freshwater Biology *49*: 681-689.

Grove et al., 2000: Grove, M. W., D. M. Finelli, D. S. Wethey & S. A. Woodin, 2000. The effects of symiotic crabs on the pumping activity and growth rates of *Chaetopterus variopedatus*, Journal of Experimental Marine Biology and Ecology *246*: 31-52.

Gunnars et al., 2002: Gunnars, A., S. Blomqvist, P. Johansson & C. Andersson, 2002. Formation of Fe(III) oxyhydroxide colloids in freshwater and brackish seawater, with incorporation of phosphate and calcium, Geochimica et Cosmochimica Acta *66*: 745-758.

Gust & Harrison, 1891: Gust, G.& J. T. Harrison, 1981. Biological pumps at the sediment-water interface: Mechanistic evaluation of the alpheid shrimp *Alpheus mackayi* and its irrigation pattern, Marine Biology *64*: 71-78.

Haglund et al., 2003: Haglund, A.-L., P. Lantz, E. Törnblom & L. Tranvik, 2003. Depth distribution of active bacteria and bacteral activity in lake sediment, FEMS Microbiology Ecology *46*: 31-38.

Hamburger et al., 1995: Hamburger, K., P. C. Dall & C. Lindegaard, 1995. Effects of oxygen deficiency on survival and glycogen content of *Chironomus anthraciunus* (Diptera, Chironomidae) under laboratory and field conditions, Hydrobiologia *297*: 187-200.

Hamer & Karius, 2005: Hamer, K.& V. Karius, 2005. Tributyltin release from harbour sediments - Modelling the influence of sedimentation, bio-irrigation and diffusion using data from Bremerhaven, Marine Pollution Bulletin *50*: 980-992.

Hammond et al., 1985: Hammond, D. E., C. Fuller, D. Harmon, B. Harman, M. Korosc, L. G. Miller, R. Rea, S. Warren, W. Berelson, & S. W. Hager, 1985. Benthic fluxes in San Francisco Bay, Hydrobiologia *129*: 69-90.

Hansen et al., 1998: Hansen, K., S. Mouridsen & E. Kristensen, 1998. The impact of *Chironomus plumosus* larvae on organic matter decay and nutrient (N, P) exchange in a shallow eutrophic lake sediment following a phytoplankton sedimentation, Hydrobiologia *364*: 65-74.

Heilskov & Holmer, 2001: Heilskov, A. C.& M. Holmer, 2001. Effects of benthic fauna on organic matter mineralization in fish-farm sediments: importance of size and abundance, Journal of Marine Science *58*: 427-434.

Helson et al., 2006: Helson, J., D. Williams & D. Turner, 2006. Larval chironomid community organization in four tropical rivers: Human impacts and longitudinal zonation, Hydrobiologia *559*: 413-431.

Hesslein, 1976: Hesslein, R. H., 1976. An in situ sampler for close interval pore water studies, Limnology and Oceanography *21*: 912-914.

Hölker & Stief, 2005: Hölker, F. & P. Stief, 2005. Adaptive behaviour of chironomid larvae (*Chironomus riparius*) in response to chemical stiumuli from predators and resource density, Behavioral Ecology and Sociobiology *58*: 256-263.

Hoppe, 1983: Hoppe, H. G., 1983. Significance of exoenzymatic activities in the ecology of brackisch water: measurements by means of methylumbelliferyl-substrates, Marine Ecology Progress Series *11*: 299-308.

Huettel, 1990: Huettel, M., 1990. Influence of the lugworm *Arenicola marina* on porewater nu-trient profiles of sand flat sediments, Marine Ecology Progress Series *62*: 241-8.

Huettel et al., 2003: Huettel, M., H. Roy, E. Precht & S. Ehrenhauss, 2003. Hydrodynamical impact on biogeochemical processes in aquatic sediments, Hydrobiologia *494*: 231-236.

Huettel & Webster 2001: Huettel, M.& I. T. Webster, 2001. Porewater flow in permeable sediments. In Boudreau, B. P. & B. B. Jorgensen (eds), The benthic boundary layer. Oxford University Press, New York: 144-179.

Hupfer et al., 2007: Hupfer, M., S. Gloess & H.-P. Grossart, 2007. Polyphosphate accumulating microorganisms in aquatic sediments, Microbial Ecology 47: 299-311.

Hupfer et al., 1995: Hupfer, M., R. Gächter & R. Giovanoli, 1995. Transformation of phosphorus species in settling seston and during early sediment diagenesis, Aquatic Sciences 57: 305-324.

Hupfer & Uhlmann, 1991: Hupfer, M.& D. Uhlmann, 1991. Microbially mediated phosphorus exchange across the mud-water interface, Verhandlungen des Internationalen Verein Limnologie 24: 2999-3003.

Huges et al., 1994: Huges, D. J., A. D. Ansell & R. J. A. Atkinson, 1994. Resource utilization by a sedentary surface deposit feeder, the echiuran worm *Maxmuelleria lankesteri*, Marine Ecology Progress Series 112: 267-275.

Joergensen & Bak, 1990: Joergensen, B. B.& F. Bak, 1990. Pathways and microbiology of thiosulfate transformations and sulfate reduction in a marine sediment (Kattegat, Denmark), Applied and Environmental Microbiology 57: 847-856.

Johnson, 1987a: Johnson, R. K., 1987a. Feeding efficiencies of *Chironomus plumosus* (L.) and *C. anthracinus* Zett. (Diptera: Chrionomedae) in mesotrophic Lake Erken, Freshwater Biology 15: 605-612.

Johnson 1987b: Johnson, R.K., 1987b. Seasonal variation in diet of *Chironomus plumosus* (L.) and *C. anthracinus* Zett. (Diptera: Chironomidae) in mesotrophic Lake Erken, Freshwater Biology 17: 525-532.

Johnson et al., 1989: Johnson, R. K., B. Boström & W. van de Bund, 1989. Interactions between *Chironomus plumosus* (L.) and the microbial community in surficial sediments of a shallow, eutrophic lake, Limnology and Oceanography 34: 992-1003.

Jones & Browser, 1978: Jones, B. F.& C. J. Browser, 1978. The mineralogy and related chemistry of lake sediments. In Lerman, A. (ed), Lake-chemistry, geology, physics, Springer Verlag. New York: 179-235.

Jorgensen et al., 1986: Jorgensen, C. B., F. Mohlenberg & O. Sten-Knudsen, 1986. Nature of relation between ventilation and oxygen consumption in filter feeders, Marine Ecology Progress Series 29: 73-88.

Kajak, 1997: Kajak, Z., 1997. *Chironomus plumosus* - what regulates its abundance in a shallow reservoir? Hydrobiologia 342/343: 133-142.

Kajak & Prus, 2004: Kajak, Z. & Prus, P., 2004. Influence of the population density and the amount of food on *Chironomus plumosus* (L.) and tubificidae. Laboratory experiments, Polish Journal of Ecology 52: 47-53.

Kalson et al., 2007: Kalson, K., E. Bonsdroff & R. Rosenberg, 2007. The impact of benthic macrofauna for nutrient fluxes from Baltic Sea sediments, Ambio (Royal Swedish Academy of Science) 36: 161-167.

Kao et al., , 2010: Kao, C.-C., J. Chen, G.-F. Chen & K. Soog, 2010. Variable swarming time of an intertidal midge (Pontomyia oceana Tokunaga, 1964) controlled by a circadian clock and temperature, Marine and Freshwater Behaviour and Physiology 43: 1-9.

Karl, 2000: Karl, D. M., 2000. Aquatic ecology: Phosphorus, the staff of life, Nature 406: 31-33.

Khalili et al., 1998: Khalili, A., A. J. Basu & U. Pietrzyk, 1998. Flow visualization in porous media via Positron Emission Tomography, Physics of Fluids 10: 1031-1033.

Khalili et al., 2000: Khalili, A., M. Huettel & W. Merzkirch, 2000. Fine-scale flow measurements in the benthic boundary layer. In Boudreau, B. P. (ed), The Benthic Boundary Layer: Transport Processes and Biogeochemistry. Oxford University Press, Oxford: 44-77

Kirilin, 2010: Kirillin, G., 2010. Modeling the impact of global warming on water temperature and seasonal mixing regimes in small temperate lakes, Boreal Environment Research 15: 279-293.

Kißner et al., 2004: Kißner, T., H. W. Riss, Stief, P. & E. I. Meyer, 2004. Bioturbations- und Bauverhalten von Chironomus plumosus (Diptera: Chironomidae) unter verschiedenen Sauerstoffbedingungen, Deutsche Gesellschaft für Limnologie (DGL), Tagungsbericht 2003 (Köln), Werder: 486-489.

Koretsky et al., 2002: Koretsky, C. M., C. Meile & P. Van Cappellen, 2002. Quantifying bioirrigation using ecological parameters: a stochastic approach, Geochemical Transactions 3: 17-30.

Kozerski & Kleeberg, 1998: Kozerski, H.-P.& A. Kleeberg, 1998. The Sediments and Benthic-Pelagic Exchange in the Shallow Lake Müggelsee (Berlin, Germany), International Review of Hydrobiology 83: 77-112.

Kristensen, 1981: Kristensen, E., 1981. Direct measurement of ventilation and oxygen uptake in three species of tubicolous polychaetes (Nereis spp.), Journal of Comparative Physiology 145: 45-50.

Kristensen, 1983a: Kristensen, E., 1983a. Ventilation and oxygen uptake by three species of Nereis (Annelida: Polychaeta). I. Effects of hypoxia, Marine Ecology Progress Series 12: 289-297.

Kristensen 1983b: Kristensen, E., 1983b. Ventilation and oxygen uptake by three species of Nereis (Annelida: Polychaeta). II. Effects of temperature and salinity changes, Marine Ecology Progress Series 12: 299-306.

Kristensen 1983c: Kristensen, E., 1983c. Comparison of polycahete (Nereis spp.) ventilation in plastic tubes and natural sediment. Marine Ecology Progress Series 12: 307–309.

Kristensen, 1988: Kristensen, E., 1988. Benthic fauna and biogeochemical processes in marine sediments: microbial activities and fluxes. In T. H. Blackburn & J. Sørensen (eds). Nitrogen cycling in coastal marine environments. John Wiley and Sons, Chichester, UK: 275–299

Kristensen, 2000: Kristensen, E., 2000. Organic matter diagenesis at the oxic/anoxic interface in coastal marine sediments, with emphasis on the role of burrowing animals, Hydrobiologia *426*: 1-24.

Kristensen, 2001: Kristensen, E., 2001. Impact of polychaetes (*Nereis* spp. and *Arenicola marina*) on carbon biogeochemistry in coastal marine sediments, Geochemical Transactions *12*: 92-103..

Kristensen & Aller, 1991: Kristensen, E.& R. C. Aller, 1991. Oxic and anoxic decomposition of tubes from the burrowing sea anemone *Ceriantheopsis americanus*: Implications for bulk sediment carbon and nitrogen balance, Journal of Marine Research *49*: 589-617.

Kristensen et al., 2009: Kristensen, E., M. H. Jensen & R. C. Aller, 2009. Direct measurement of dissolved inorgganic nitrogen exchange and denitrification in individual polychaete (*Nereis virens*) burrows, Journal of Marine Research *49*: 355-377.

Kristensen et al., 1985: Kristensen, E., M. H. Jensen & T. K. Andersen, 1985. The impact of polychaete (*Nereis virens* Sars) burrows on nitrification and nitrate reduction in estuarine sediments, Journal of Experimental Marine Biology and Ecology *85*: 75-91.

Kristensen & Kostka, 2005: Kristensen, E.& J. E. Kostka, 2005. Macrofaunal burrows and irrigation in marine sediment: Microbiological and biogeochemical interactions, The Ecogeomorphology of Tidal Marshes *59*: 1-34.

Kristensen & Mikkelsen, 2003: Kristensen, E.& O. L. Mikkelsen, 2003. Impact of the burrow-dwelling polychaete *Nereis diversicolor* on the degradation of fresh and aged macroalgal detritus in a coastal marine sediment, Marine Ecology-Progress Series *265*: 141-153.

Krom, 1980: Krom, M. D., 1980. Spectrophotometric determination of ammonia: A study of a modified Berthelot reaction using salicylate and dichloroisocyanurate, Analyst *105*: 305-316.

Krom et al., 1994: Krom, M. D., P. Davison, H. Zhang & W. Davison, 1994. High-resolution pore-water sampling with a gel sampler, Limnology and Oceanography *39*: 1967-1972.

Krüger, 1964: Krüger, F., 1964. Messung der Pumpaktivität von *Arenicola marina* L. im Watt, Helgoländer wissenschaftliche Meeresuntersuchungen *11*: 70-91.

Krüger, 1971: Krüger, F., 1971. Bau und Leben des Wattwurmes *Arenicola marina,* Helgoland Marine Research *22*: 149-200.

Kühl & Revsbech, 2001: Kühl, M.& N. P. Revsbech, 2001. Biogeochemical microsensors for boundary layer studies. In Boudreau, B. P. & B. B. Jorgensen (eds), The benthic boundary layer: Transport processes and biogeochemistry. Oxford University Press, New York: 180-211.

Langton & Pinder, 2007a: Langton, P. H., & L. C. V. Pinder, 2007a. Keys to the adult male Chironomidae of Britain and Ireland. Volume 1. Freshwater Biological Association, Cumbria, UK.

Langton & Pinder, 2007b: Langton, P. H., & L. C. V Pinder, 2007b. Keys to the adult male Chironomidae of Britain and Ireland. Volume 2. Freshwater Biological Association, Cumbria, UK.

Laskov et al., 2006: Laskov, C., C. Herzog, J. Lewandowski & M. Hupfer, 2006. Miniaturized photometrical methods for the rapid analysis of phosphate, ammonium, ferrous iron, and sulfate in pore water of freshwater sediments, Limnology and Oceanography: Methods 4: 63-71.

Lauer, 1969: Lauer, G. J.,1969. Osmotic regulation of Tanypus nubifer, Chironomus plumosus and Enalagma clausum in various concentrations of saline lake water. Physiology and Zoology 42: 381-387.

Laverock et al., 2010: Laverock, B., C. J. Smith, K. Tait, A. M. Osborn, S. Widdecombe & J. A. Gilber, 2010. Bioturbating shrimp alter the structure and diversity of bacterial communities in coastal marine sediments, The ISME Journal 4: 1531-1544

Lee et al., 1986: Lee, T. M. & M. K. McClintock, 1986. Female rats in a laboratory display seasonal-variation in fecundity. Journal of Reproduction and Fertility 77: 51-59.

Lehours et al., 2007: Lehours, A.-C., P. Evans, C. Bardot, K. Joblin & F. Gérard, 2007. Phylogenetic diversity of archaea and bacteria in the anoxic zone of a meromictic lake (Lake Pavin, France), Applied and Environmental Microbiology 73: 2016-2019.

Leuchs, 1986: Leuchs, H., 1986. The ventilation activity of *Chironomus larvae* (Diptera) from shallow and deep lakes and the resulting water circulation in correlation to temperature and oxygen conditions, Archiv für Hydrobiologie 108: 281-299.

Leuchs & Neumann, 1990: Leuchs, H.& D. Neumann, 1990. Tube texture, spinning and feeding behaviour of *Chrionomus* larvae, Zoologisches Jahrbuch für Systematik 117: 31-40.

Lewandowski, 2002: Lewandowski, J., 2002. Untersuchungen zum Einfluss seeinterner Verfahren auf die Phosphor-Diagenese in Sedimenten. Institut für Gewässerökolgie und Binnenfischerei (IGB), Dissertation, Berlin.

Lewandowski & Hupfer, 2005a: Lewandowski, J.& M. Hupfer, 2005a. Effect of macrozoobenthos on two-dimensional small-scale heterogeneity of pore water phosphorus concentrations in lake sediments: A laboratory study, Limnology and Oceanography 50: 1106-1118.

Lewandowski & Hupfer, 2005b: Lewandowski, J.& M. Hupfer, 2005b. Impact of macrozoobenthos on two-dimensional small-scale heterogeneity of pore water phosphorus concentration in lake sediments. In Serrano, L. & H. L. Golterman (eds), Phosphates in Sediments. Backhuys Publishers, Lleiden: 171-172.

Lewandowski et al., 2007: Lewandowski, J., C. Laskov & M. Hupfer, 2007. The relationship between *Chironomus plumosus* burrows and the spatial distribution of pore-water phosphate, iron and ammonium in lake sediments, Freshwater Biology 52: 331-343.

Lewandowski et al., 2002: Lewandowski, J., S. Rütger & M. Hupfer, 2002. Two-dimensional small-scale variability of pore water phosphate in freshwater lakes: results from a novel dialysis sampler, Environmental Science Technology 36: 2039-2047.

Lewandowski et al., 2006: Lewandowski, J., M. Schadach & M. Hupfer, 2006. Impact of macrozoobenthos on two-dimensional small-scale heterogeneity of pore water phosphorus concentrations: in-situ study Lake Arendsee (Germany), Hydrobiologia 549: 43-55.

Luther et al., 2008: Luther III, G. W., B. T. Glazer, S. Ma, R. E. Roubworst, T. S. Moore, E. Mezger, C. Kraiya, T. Waite, G. K. Druschel, B. Sundby, M. Teillefert, D. B. Nuzzio, T. M. Shank & B. Lewis, 2008. Use of voltammetric solid-state (micro)electrodes for studying biogeochemical processes: from laboratory measurements to real time measurements with an in situ electrochemical analyzer (ISEA), Marine Chemistry 108: 221-235.

Martin et al., 1987: Martin, W. R.& F. L. Sayles, 1987. Seasonal cycles of particle and solute transport processes in nearshore sediments - $^{222}Rn/^{226}Ra$ and $^{234}Th/^{238}U$ disequilibrium at a site in Buzzards Bay, Ma, Geochimica et Cosmochimica Acta 51: 927-943.

Marxen, 1996: Marxen, J., 1996. Measurement of bacterial production in stream-bed sediments via leucine incorporation, FEMS Microbiology Ecology 21: 313-325.

Marxen et al., 1998: Marxsen, J., P. Tippmann, P. Heininger, P. Preuß & A. Remde, 1998. Aktivität - Enzyme - Enzymaktivität. In Remde, A. & P. Tippmann (eds), Mikrobiolgische Chrarakterisierung aquatischer Sedimente - Methodensammlung -. Vereinigung für Allgemeine und Angewandte Mikrobiologie (VAAM), R. Oldenbourg Verlag, München Wien: 87-109.

Matisoff et al., 1985: Matisoff, G., J. B. Fisher & S. Matis, 1985. Effects of benthic macroinvertebrates on the exchange of solutes between sediments and freshwater, Hydrobiologia 122: 19-33.

Mayer et al., 1995: Mayer, M. S., L. Schaffner & W. M. Kimp, 1995. Nitrification potentials of benthic macrofaunal tubes and burrow walls: effects of sediment NH_4^+ and animal irrigation behavior, Marine Ecology Progress Series 121: 157-169.

McLachlan, 1977: McLachlan, A. J., 1977. Some effects of tube shape on the feeding of Chironomus plumosus L. (Diptera: Chironomidae), Journal of Animal Ecology 46: 139-146.

McLachlan & Cantrell, 1976: McLachlan, A. J. & M. A. Cantrell, 1976. Sediment development and its influence on the dirstibution and tube structure of Chironomus plumosus L. (Chironomidae, Diptera) in a new impoundment, Freshwater Biology 6: 437-443.

Meile et al., 2001: Meile C., C.M. Koretsky & P. Van Cappellen, 2001. Quantifying bioirrigation in aquatic sediments: An inverse moldeling approach. Limnology and Oceanogrphy 46: 164-77.

Meile et al., 2003: Meile, C., K. Tuncay & P. Van Capellen, 2003. Explicit representation of spatial geterogeneity in reactive transport models: application to bioirrigated sediments, Journal of Geochemical Exploration 78-79: 231-234.

Mermillod-Blondin et al., 2008: Mermillod-Blondin, F., D. Lemoine, J.-C. Boisson, E. Malet & B. Montuelle, 2008. Relative influences of submersed macrophytes and bioturbating fauna on biogeochemical processes and microbial activities in freshwater sediments, Freshwater Biology 53: 1969-1982.

Mermillod-Blondin et al., 2005: Mermillod-Blondin, F., G. Nogaro, T. Datry, F. Malard & J. Gibert, 2005. Do tubificid worms influence the fate of organic matter and pollutants in stormwater sediments?, Environmental Pollution 134: 57-69.

Meyer-Reil, 1987: Meyer-Reil, L.-A., 1987. Seasonal and spatial distribution of extracellular enzymatic activities and microbial lincorporation of dissolved organic substrates in marine sediments, Applied and Environmental Microbiology 53: 1748-1755.

Meysman et al., 2006a: Meysman, F. J. R., O. S. Galaktionov, B. Gribsholt & J. J. Middelburg, 2006a. Bioirrigation in permeable sediments: Advective pore-water transport induced by burrow ventilation, Limnology and Oceanography 51: 142-156.

Meysman et al., 2006b: Meysman, F. J. R., O. S. Galaktionov, B. Gribsholt & J. J. Middelburg, 2006b. Bio-irrigation in permeable sediments: An assessment of model complexity, Journal of Marine Research 64: 589-627.

Mortimer, 1941: Mortimer, C. H., 1941. The exchange of dissolved substances between mud and water in lakes. Journal of Ecology, 29: 280-329.

Murphy & Riley, 1962: Murphy, J.& J. P. Riley, 1962. A modified single solution method for the determination of phosphate in natural waters, Analytica Chimica Acta 27: 31-36.

Mutz et al., 2007: Mutz, M., E. Kalbus & S. Meinecke, 2007. Effect of instream wood on vertical water flux in low-energy sand bed flume experiments, Water Resources Research 43: 1-10.

Na et al., 2008: Na, T., B. Gribsholt, O. S. Galaktionov, T. Lee & F. J. R. Meysman, 2008. Influence of advective bio-irrigation on nitrogen cycling in sandy sediment, Journal of Marine Research 66: 691-722.

Nagell & Landahl, 1978: Nagell, B. & C. C. Landahl, 1978. Resistance to anoxia of *Chironomus plumosus* and *Chrionoumus anthracinus* (Diptera) larvae, Holarctic Ecology 1: 333-336.

Naylor, 1963: Naylor, E., 1963. Temperature relationships of the locomotor rhythm of Carcinus. Journal of Experimental Biology 40: 669-679.

Neumann, 1965: Neumann, D., 1965. Photoperiodische Steuerung der 15-tägigen lunaren Metamorphose-Periodik von *Clunio* Population (Diptera: Chironomidae), Zeitschrift für Naturforschung 206: 818-819.

Nielsen et al., 1995: Nielsen, A. M., N. T. Eriksen, J. J. L. Iversen & H. U.Riisgård, 1995. Feeding, growth and respiration in the polychaetes *Nereis diversicolor* (facultative filter-feeder) and *N. virens* (omnivorous) – a comparative study, Marine Ecology Progress Series 125: 149–158.

Nixdorf, 1004: Nixdorf, B., 1994. Polymixis of a shallow lake (Großer Müggelsee, Berlin) and its influence on seasonal phytoplankton dynamics, Hydrobiologia 275/276: 173–186.

Nogaro et al., 2007: Nogaro, G., F. Mermillod-Blondini, B. Montuelle, J.-C. Boisson, J.-F. Bedell, A. Ohannessian, B. Volat & J. Gibert, 2007. Influence of a stormwater sediment deposit on

Oliver, 1971: Oliver, D. R., 1971. Life history of the Chironomidae, Annual Review of Entomology 16: 211–230.

Osovitz & Julian, 2002: Osovitz, C. J.& D. Julian, 2002. Burrow irrigation behaviour of *Urechis caupo*, a filter-feeding marine invertebrate, in its natural habitat, Marine Ecology Progress Series 245: 149-155.

Papaspyrou et al., 2005: Papaspyrou, S., T. Gregersen, R. P. Cox, M. Thessalou-Legaki & E. Kristensen, 2005. Sediment properties and bacterial community in the burrows of the mud shrimp *Pestarella tyrrhena* (Decapoda: Thalassinidea), Aquatic Microbial Ecology 38: 181-190.

Papaspyrou et al., 2006: Papaspyrou, S., T. Gregersen, E. Kristensen, B. Christensen & R. P. Cox, 2006. Microbial reaction rates and bacterial communities in sediment surrounding burrows of two nereidid polychaetes (*Nereis diversicolor* and *N. virens*), Marine Biology 148: 541-550.

Pelegri & Blackburn, 1995: Pelegri, S. P.& T. H. Blackburn, 1995. Effects of *Tubifex tubifex* (Oligochaeta: Tubificidae) on N-mineralization in freshwater sediments, measured with ^{15}N isotopes, Aquatic Microbial Ecology 9: 289-294.

Pelegri & Blackburn, 1996: Pelegri, S. P.& T. H. Blackburn, 1996. Nitrogen cycling in lake sediments bioturbated by *Chironomus plumosus* larvae, under different degrees of oxygenation, Hydrobiologia 325: 231-238.

Pelegri et al., 1994: Pelegri, S. P., L. P. Nielsen & T. H. Blackburn, 1994. Denitrification in eustarine sediment stimulated by the irrigation activity of the amphipod *Corophium voutator*, Marine Ecology Progress Series 105: 285-290.

Pinder, 1986: Pinder, L., 1986. Biology of freshwater Chironomidae, Annual Review of Entomology 31: 1-23.

Plazer, 1967: Platzer, I., 1967. Untersuchungen zur Temperaturadaption der tropischen Chironomidenart *Chironomus streskei* Fittkau (Dipters), Zeitschrift für vergleichende Physiologie 54: 58-74.

Polerecky et al., 2005: Polerecky, L., U. Franke, U. Werner, B. Grunwald & D. de Beer, 2005. High spatial resolution measurement of oxygen consumption rates in permeable sediments, Limnology and Oceanography: Methods 3: 75-85.

Polerecky et al., 2006: Polerecky, L., N. Volkenborn & P. Stief, 2006. High temporal resolution oxygen imaging in bioirrigated sediments, Environmental Science Technology 40: 5763-5769.

Porter & Geig, 1980: Porter, K. G.& Y. S. Feig, 1980. The use of DAPI for identifying and counting aquatic microflora, Limnology and Oceanography 25: 943-948.

Precht et al., 2004: Precht, E., U. Franke, L. Polerecky & M. Huettel, 2004. Oxygen dynamics in permeable sediments with wave-driven pore water exchange, Limnology and Oceanography 49: 693-705.

Psenner et al., 1988: Psenner, R., B. Boström, M. Dinka, K. Petterson, R. Pucsko & M. Sager, 1988. Fractionation of phosphorus in suspended matter and sediment. Archiv für Hydrobiologie Beiheft Ergebnis Limnologie *30*: 98-103.

Psenner et al., 1984: Psenner, R., R. Pucsko & M. Sager, 1984. Fractionation of organic and inorganic phosphorus compounds in lake sediments, Archiv für Hydrobiologie *70*: 111-155.

Pusch et al., 2001: Pusch, M., J. Liefert & N. Walz, 2001. Filtration and respiration rates of two unionid species and their impact on the water quality of a lowland river. In Bauer, G. W. K. (ed), Ecology and Evolution of the Freshwater Mussels Unionoida. Springer, Berlin: 317-326.

Rasmussen et al., 1998: Rasmussen, A. D., G. T. Banta & O. Andersen, 1998. Effects of bioturbation by the lugworm *Arenicola marina* on cadmium uptake and distribution in sandy sediments, Marine Ecology Progress Series *164*: 179-188.

Reichardt, 1988: Reichardt, W., 1988. Impacto of bioturbation by *Arenicola marina* on microbiological parameters in intertidal sediments, Marine Ecology *44*: 149-158.

Reiche et al., 2009: Reiche, M., A. Hädrich, G. Lischeid & K. Küsel, 2009. Impact of manipulated drought and heaby rainfall events on peat minaeralization porcesses and source-sink functions of an acidic fen, Journal of Geophysical Research *114*: 1-13.

Revsbech et al., 1998: Revsbech, N. P., L. P. Nielsen & N. B. Ramsing, 1998. A novel microsensor for determination of apparent diffusivity in sediments, Limnology and Oceanography *43*: 986-992.

Revsbech 1989: Revsbech, N. P., 1989. An O2 microelectrode with a guard cathode, Limnology and Oceanography *34*: 472-476.

Riisgard, 1991: Riisgard, H. U., 1991. Suspension feeding in the polychaete *Nereis diversicolor*, Marine Ecology Progress Series *70*: 29-37.

Riisgard, 2007: Riisgard, H. U., 2007. Biomechanics and energy cost of the amphipod *Corophium volutato*r filter-pump, Biological Bulletin *212*: 104-114.

Riisgard & Banta, 1998: Riisgard, H. U.& G. T. Banta, 1998. Irrigation and deposit feeding by the lugworm *Arenicola marina*, characteristics and secondary effects on the environment. A review of current knowledge, Vie Milieu *48*: 243-257.

Riisgard & Larsen, 2001: Riisgard, H. U.& P. S. Larsen, 2001. Minireview: Ciliary filter feeding and bio-fluid mechanics - present understanding and unsolved problems, Limnology and Oceanography *46*: 882-891.

Riisgard & Larsen, 2005: Riisgard, H. U.& P. S. Larsen, 2005. Water pumping and analysis of flow in burrowing zoobenthos: an overview, Aquatic Ecology *39*: 237-258.

Riisgard et al., 1992: Riisgard, H. U., A. Vedel, H. Boye & P. S. Larsen, 1992. Filter-net structure and pumping activity in the polychaete *Nereis diversicolor*: effects of temperature and pump-modelling, Maine Ecology Progress Series *83*: 79-89.

Rosenberg, 2001: Rosenberg, R., 2001. Marine benthic faunal successional stages and related sedimentary activity, Scientia Marina *65*: 107-119.

Roskosch et al., 2011: Roskosch, A., M. Hupfer, G. Nützmann, & J. Lewandowski, 2011. Measurement techniques for quantification of pumping activity of invertebrates in small burrows, Fundamental and Applied Limnology *178*: 89-110.

Roskosch et al., 2010: Roskosch, A., J. Lewandowski, R. Bergmann, F. Wilke, W. Brenner, & R. Buchert, 2010. Identification of transport processes in bioirrigated muddy sediments by [18F]fluoride PET (Positron Emission Tomography), Applied Radiation and Isotopes *68*: 1094-1097.

Roskosch et al., 2010: Roskosch, A., M. R. Morad, A. Khalili, & J. Lewandowski, 2010. Bioirrigation by Chironomus plumosus: advective flow investigated by particle image velocimetry, Journal of the North American Benthological Society *29*: 789-802.

Saunders et al., 2004: Saunders, D. S., R. D. Lewis & G. R. Warman, 2004. Photoperiodic induction of diapause: opening the black box, Physiological Entomology *29*: 1-15.

Schindler, 1987: Schindler, D. W., 1978. Factors regulating phytoplankton production and standing crop in the world's freshwaters, Limnology and Oceanography *23*: 478-786.

Schlüter et al., 2000: Schlüter, M., E. Sauter, H.-P. Hansen & E. Suess, 2000. Seasonal variations of bioirrigation in coastal sediments: Modelling of field data, Geochimica et Cosmochimica Acta *64*: 821-834.

Schreiber et al., 2009: Schreiber, F., B. Loeffler, L. Polerecky, M. M. M. Kuypers & D. de Beer, 2009. Mechanisms of transient nitric oxide and nitrous oxide production in a complex biofilm, The ISME Journal *3*: 1301-1313.

Seeberg-Elverfeldt et al., 2005: Seeberg-Elverfeldt, J., M. Schlüter, T. Feseker & Kölling M., 2005. Rhizon sampling of porewaters near the sediment-water interface of aquatic systems, Limnology and Oceanography: Methods *3*: 361-371.

Seymour, 1972: Seymour, M. K., 1972. Effects of temperature change on irrigation rate in *Arenicola marina* (L.), Comparative Biochemistry and Physiology Part A: Molecular & Integrative Physiology *43*: 553-564.

Shpigel, 2005: Shpigel, M., 2005. Bivalves as biofilters and valuable byproducts in land-based aquaculture systems. In Dame, R. and Olenin, S. (eds), The Comparative Roles of Suspension-Feeders in Ecosystems, Springer, Dordrecht, The Netherlands: 183-197.

Shull, 2001: Shull, D. H., 2001. Bioturbation. In Steele, J., S. Thorpe & B. Turekian (eds), Encyclopedia of Ocean Sciences. Elsevier LTD, Academic Press, Oxford: 1-6.

Shull et al., 2009: Shull, D.H., J. M. Benoit, C. Wojcik & J. R. Senning, 2009: Infaunal burrow ventilation and pore-water transport in muddy sediments, Coastal and Shelf Science *83*: 277-286.

Smethie et al., 2003: Smethie, W. M., C. A. Nittrouer & R. F. L. Self, 2003. The use of radon-222 as a tracer of sediment irrigation and mixing on the Washington continental shelf, Marine Geology *42*: 173-200.

Sørensen et al., 1979: Sørensen, J., B. B. Jargensen & N. P., Revsbeck, 1979. A comparison of oxygen, nitrate, and sulfate respiration in coastal marine sediments, Journal of Microbial Ecolology 5: 105-115.

Stamhuis et al., 1996: Stamhuis, E. J., T. Reede-Dekker, Y. van Etten, J. J. de Wiljes & J. J. Videler, 1996. Behaviour and time allocation of the burrowing shrimp *Callianassa subterranea* (Decapoda, Thalassinidea), Journal of Experimental Marine Biology and Ecology 204: 225-239.

Stamhuis & Videler, 1995: Stamhuis, E. J. & J. J. Videler, 1995. Quantitative flow analysis around aquatic animals using laser sheet Particle Image Velocimetry, Journal of Experimental Biology 198: 283–294.

Stamhuis & Videler, 1998: Stamhuis, E. J.& J. J. Videler, 1998. Burrow ventilation in the tube-dwelling shrimp *Callianassa subterranea* (Decapoda: thalassinidea). II. The flow in the vicinity of the shrimp and the energetic advantages of a laminar non-pulsating ventilation current, The Journal of Experimental Biology 201: 2159-2170.

Stewerd et al., 1996: Steward, C. C., S. C. Nold, D. B. Ringelberg, D. C. White & C. R. Lovell, 1996. Micobial biomass and community structures in the burrows of bromophenol producing and non-producing marine worms and surrounding sediments, Marine Ecology Progress Series 133: 149-165.

Stief, 2007: Stief, P., 2007. Enhanced exoenzyme activities in sediments in the presence of deposit-feeding *Chironomus riparius* larvae, Freshwater Biology 52: 1807-1819.

Stief et al., 2004: Stief, P., D. Altmann, D. de Beer, R. Bieg & A. Kureck, 2004. Microbial activities in the burrow environment of the potamal mayfly *Ephoron virgo*, Freshwater Biology 49: 1152-1163.

Stief & de Beer, 2002: Stief, P.& D. de Beer, 2002. Bioturbation effects of *Chironomus riparius* on the benthic N-cycle as measured using microsensors and microbiological assys, Aquatic Micorbial Ecology 27: 175-185.

Stief & de Beer, 2006: Stief, P.& D. de Beer, 2006. Probing the microenvironment of freshwater sediment macrofauna: Implications of deposit-feeding and bioirrigation for nitrogen cycling, Limnology and Oceanography 51: 2538-2548.

Stief & Hölker, 2006: Stief, P.& F. Hölker, 2006. Trait-mediated indirect effects of predatory fish on microbial mineralization in aquatic sediments, Ecological Archives 87: 3152-3159.

Stief et al., 2009: Stief, P., M. Poulsen, L. P. Nielsen, H. Brix & A. Schramm, 2009. Nitrous oxide emission by aquatic macrofauna, PNAS 106: 4296-4300.

Stief & Schramm, 2010: Stief, P.& A. Schramm, 2010. Regulation of nitrous oxide emission associated with benthic invertebrates, Freshwater Biology 55: 1647-1657.

Strathmann, 1971: Strathmann, R. R., 1971. The feeding behavior of planktotrophic echinoderm larvae: mechanisms, regulation and rates of suspension-feeding, Journal of Experimental Marine Biology and Ecology 6: 109–160.

Svensson, 1997: Svensson, J. M., 1997. Influence of *Chironomus plumosus* larvae on ammonium flux and denitrification (measured by the acetylene blockage- and the isotope pairing-technique) in eutrophic lake sediment, Hydrobiologia *346*: 157-168.

Svensson & Leonardson, 1996: Svensson, J. M.& L. Leonardson, 1996. Effects of bioturbation by tube-dwelling chironomid larvae on oxygen uptake and denitrification in eutrophic lake sediments, Freshwater Biology *35*: 289-300.

Tabatabai, 1974: Tabatabai, M. A., 1974. A rapid method for determination of sulfate in water samples, Environmental Letters *7:* 237-243.

Tamura et al., 1974: Tamura, H., K. Goto, T. Yotsuyanagi & M. Nagayama, 1974. Spectophotometric determination of iron(II) with 1,10-phenanthroline in the presence of large amounts of rion(III), Talanta *21*: 314-318.

Teasdale et al., 2010: Teasdale, P. R., G. E. Batley, S. C. Apte & I. T. Webster, 2010. Pore water sampling with sediment peepers, Trends in Analytical Chemistry *14*: 256.

Tellioglu et al., 2008: Tellioglu, A., C. Citil & I. Sahin, 2008. Distribution of Chironomidae (Diptera) larvae in Hazar Lake, Turkey, Journal of Applied Biological Sciences *2*: 77-80.

Tezuka, 1990: Tezuka, Y., 1990. Bacterial regeneration of ammonium and phosphate as affected by carbon: nitrogen: phosphorus ratio of organic substrates, Microbial Ecology *19*: 227-238.

Thienemann, 1954: Thienemann, A., 1954. Chironomus – Leben, Verbreitung und wirtschaftliche Bedeutung der Chironomiden , 2nd edn. E. Schweizerbart´sche Verlagsbuchhandlung, Stuttgart, Germany.

Timmermann et al., 2002: Timmermann, K., J. H. Christensen & G. T. Banta, 2002. Modeling of advective solute transport in sandy sediments inhabited by the lugworm *Arenicola marina*, Journal of Marine Research *60*: 151-169.

Timmermann et al., 2006: Timmermann, K., G. T. Banta, & R. N. Glud, 2006. Linking *Arenicola marina* irrigation behavior to oxygen transport and dynamics in sandy sediments, Journal of Marine Research *64*: 915–938.

Toulmond & Dejours, 1994: Toulmond, A. & P. Dejours, 1994. Energetics of the ventilatory piston pump of lugworm, a deposit-feeding polychaete living in a burro, Biological Bulletin *186*: 213-222.

Traunspurger et al., 1997: Traunspurger, W., M. Bergold & W. Goedkoop, 1997. The effects of nematodes on bacterial activity and abundance in a freshwater sediment, Oecologia *112*: 118-122.

Van de Bund et al., 1994: Van de Bund, W. J., W. Goedkoop & R. K. Johnson, 1994. Effects of deposit-feeder activity on bacterial production and abundance in profundal lake sediment, Journal of the North American Benthological Society *13*: 532-539.

Van Dijk et al., 2005: Van Dijk, P. L. M., I. Hardewig & F. Hölker, 2005. Energy reserves during food deprivation and compensatory growth in juvenile roach: the importance of season and temperature, Journal of Fish Biology *66*: 167-181.

Van Rees et al., 1996: Van Rees, K. C. J., K. R. Reddy & P. S. C. Rao, 1996. Influence of benthic organisms on solute transport in lake sediments, Hydrobiologia *317*: 31-40.

Vedel et al., 1994: Vedel, A., B. B. Adresen & H. U. Riisgård, 1994. Field investigations of pumping activity of the facultatively filter-feeding polychaete *Nereis diversicolor* using an improved infrared phototransducer system, Marine Ecology Progress Series *103*: 91–101.

Volkenborn et al., 2007: Volkenborn, N., S. I. C. Hedtkamp, J. E. E. van Beusekom & K. Reise, 2007. Effects of bioturbation and bioirrigation by lugworms (*Arenicola marina*) on physical and chemical sediment properties and implications for intertidal habitat succession, Estuarine, Coastal and Shelf Science *74*: 331-343.

Vopel et al., 2002: Vopel, K., C. H Reick., G. Arlt, M. Pöhn & J. A. Ott, 2002: Flow microenvironment of two marine peritrich ciliates with ectobiotic chemoautotrophic bacteria, Aquatic Microbiology and Ecology *29*: 19-28.

Waldbusser & Marinelli, 2006: Waldbusser, G. G.& R. L. Marinelli, 2006. Macrofaunal modification of porewater advection: role of species function, species interaction, and kinetics, Marine Ecology Progress Series *311*: 217-231.

Wallace & Merritt, 1980: Wallace, J. B. & R. W. Merritt, 1980. Filter-feeding ecology of aquatic insects, Annual Review of Entomology *25*: 103-132.

Walshe, 1947: Walshe, B. M., 1947. Feeding mechanism of *Chironomus plumosus* larvae, Nature *160*: 474.

Walshe, 1948: Walshe, B. M., 1948. The oxygen requirements and thermal resistance of Chironomid larvae from flowing and from still waters, Journal of Experimental Biology *25*: 35-44.

Walshe, 1950a: Walshe, B. M., 1950a. The function of haemoglobin in *Chironomus plumosus* under natural conditions, Journal of Experimental Biology *27*: 73-95.

Walshe 1950b: Walshe, B. M., 1950b. The function of haemoglobin in relation to filter feeding in leaf-mining chironomid larvae, Journal of Experimental Biology *28*: 57-61.

Wang et al., 2001: Wang, F., A. Tessier & L. Hare, 2001. Oxygen measurements in the burrows of freshwater insects, Freshwater Biology *46*: 317-327.

Wang & Van Cappellen, 1996: Wang, Y.& P. Van Cappellen, 1996. A multicomponent reactive transport model of early diagenesis: Application to redox cycling in coastal marine sediments, Geochimica et Cosmochimica Acta *60*: 2993-3014.

Weaver & Schulheiss, 1983: Weaver, P. P. E.& P. J. Schultheiss, 1983. Vertical open burrows in deep-sea sediments 2 m in length, Nature *301*: 329-331.

Weber & Bauer, 2004: Weber, S. & A. Bauer, 2004. Small animal PET: aspects of performance assessment, European Journal of Nuclear Medicine and Molecular Imaging *1*: 1545-55.

Wethey & Woodin, 2005: Wethey, D. S. & S. A. Woodin, 2005. Infaunal hydraulics generate porewater pressure signals, Biological Bulletin *209*: 139-145.

Wilhelm et al., 2006: Wilhelm, S., T. Hinze, D. M. Livingstone & R. Adrian, 2006. Long-term response of daily epilimnetic temperature extrema to climate forcing, Canadian Journal of Fisheries and Aquatic Sciences 63: 2467-2477.

Wilhelm & Adrian, 2008: Wilhelm, S. & R. Adrian, 2008. Impact of summer warming on the thermal characteristics of a polymictic lake and consequences for oxygen, nutrients and phytoplankton, Freshwater Biology 53: 226-237.

Woodin et al., 2003: Woodin, S. A., R. A. Merz, F. M. Thomas, D. R. Edwards & I. L. Garcia, 2003. Chaetae and mechanical function: tools no Metazoan class should be without, Hydrobiologia 496: 253-258.

Yeager et al., 2001: Yeager, P. E., C. L. Foreman & R. L. Sinsabaugh, 2001. Microbial community structure and function in response to larval chironomid feeding pressure in a microcosm experiment, Hydrobiologia 448: 71-81.

Zhou et al., 1996: Zhou, J., M. A. Bruns & J. M. Tiedja, 1996. DNA recovery from soils of diverse composition, Applied and Environmental Microbiology 62: 316-322.

Zorn et al., 2006: Zorn, M. E., S. V. Lalonde, M. K. Gingras, S. G. Pemberton & K. O. Konhauser, 2006. Microscale oxygen distribution in various invertebrate burrow walls, Geobiology 4: 137-145.

Zorn et al., 2010: Zorn, M.E., M. K. Gingras & S.G. Pemberton, 2010. Variation in burrow-wall micromorphologies of selected intertidal invertebrates along the Pacific Northwest coast, USA: Behavioral and diagentic implications, PALAIOS. 25: 59-72.

Zwirnmann et al., 1999: Zwirnmann, E., A. Krüger & J. Gelbrecht, 1999. Analytik im Zentralen Chemielabor des IGB, IGB-Berichte, Institut für Gewässerökolgie und Binnenfischerei (IGB), Berlin, 9: 3-24.

Appendix

Appendix A: Data of the sampling of the *C. plumosus* larvae from Lake Müggelsee, Berlin (6 m water depth, ca. N 52°44´ and E 13°65´) including the temperatures of air and water. For the calculation of the arithmetic mean of larvae the samplings between November 2007 and December 2008 were taken into account (chapter II.I and II.III).

Appendix B: X-ray image of two *C. plumosus* larvae dwelling in a mesocosm. Larvae were marked with silver conductive paint (Busch, Germany); two points paints are visible in the back and one point in the front of their body. The sediment was mixed with 20 g L^{-1} Mo$_2$C. One of the larvae (right) died.

Appendix C: Results of the PET measurements. Left figure shows the Computer Tomography (CT-) image of the column with the burrow of a *C. plumosus* larva. Middle figure shows the top view of the column with burrow inlet and outlet. Right figure shows the PET-image of the column with the burrow. A deeper penetration depth at burrow outlet side than at the burrow inlet side is visible.

Appendix D: Determination of the k_f-value (coefficient of hydraulic conductivity) of the sediments of Lake Müggelsee.

Date	Larvae m^{-2} (4. instar)	Temperature air °C	Temperature water °C
29.11.2007	1.401	6.4	2.4
31.03.2008	747	14.7	6
15.05.2008	624	24	18
01.07.2008	849	24.2	21.2
15.07.2008	809	20.2	20.7
01.09.2008	292	22.1	19.0
06.10.2008	692	-	-
10.12.2008	545	2.3	2.8
10.02.2009	-	1.5	1.7
08.04.2009	480	18	11.1
02.06.2009	185	21	17
20.07.2009	533	19.9	20.9
11.09.2009	220	17.1	18.6
09.11.2009	478	4.9	6.2
26.01.2010	376	-15	0
03.02.2010	284	3.2	0
26.03.2010	527	22.6	6.5
17.05.2010	223	15.4	10.9
22.06.2010	331	19	17.6
03.08.2010	101	19.2	21.3
02.11.2010	590	9.9	6.6
arithmetic mean	514 ± 299		
arithmetric mean (11.2007 to 12.2008)	745 ± 318		

Appendix A: Data of the sampling of the *C. plumosus* larvae from Lake Müggelsee, Berlin (6 m water depth, ca. N 52°44´ and E 13°65´) including the temperatures of air and water. For the calculation of the arithmetic mean of larvae the samplings between November 2007 and December 2008 were taken into account (chapter II.I and II.III).

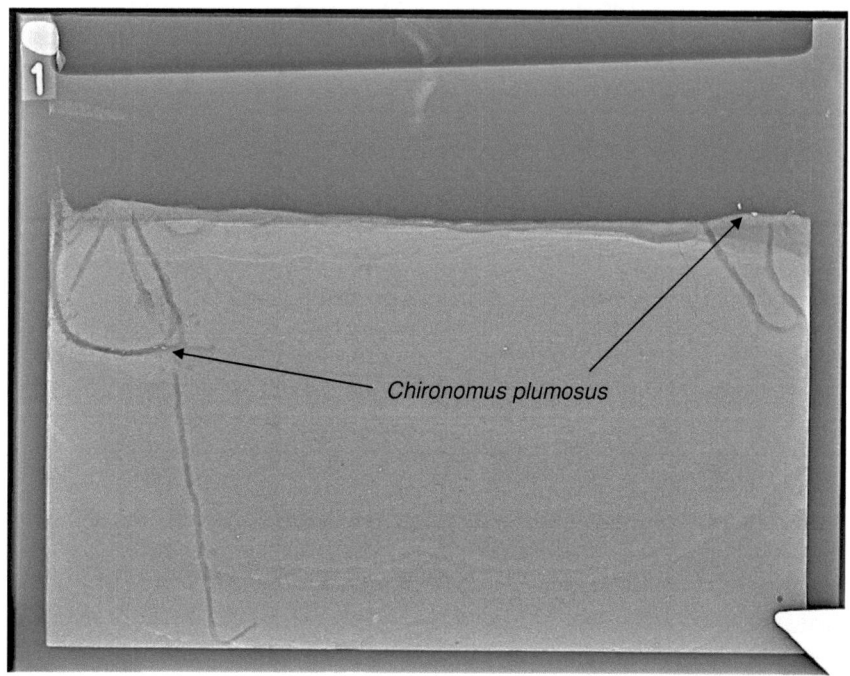

Appendix B: X-ray image of two *C. plumosus* larvae dwelling in a mesocosm. Larvae were marked with silver conductive paint (Busch, Germany); two points paints are visible in the back and one point in the front of their body. The sediment was mixed with 20 g L^{-1} Mo$_2$C. One of the larvae (right) died.

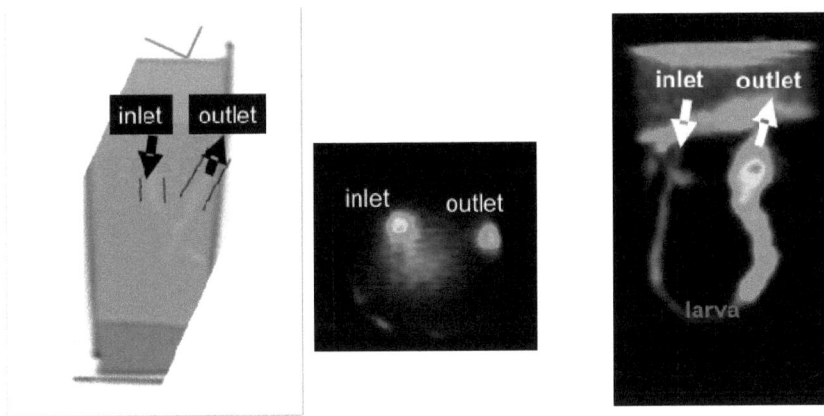

Appendix C: Results of the PET measurements. Left figure shows the Computer Tomography (CT-) image of the column with the burrow of a *C. plumosus* larva. Middle figure shows the top view of the column with burrow inlet and outlet. Right figure shows the PET-image of the column with the burrow. A deeper penetration depth at burrow outlet side than at the burrow inlet side is visible.

Experiment-no.	k_f
1	$5{,}96 \times 10^{-6}$
	$6{,}74 \times 10^{-6}$
	$6{,}29 \times 10^{-6}$
	$6{,}07 \times 10^{-6}$
	$6{,}52 \times 10^{-6}$
2	$4{,}18 \times 10^{-6}$
	$4{,}64 \times 10^{-6}$
	$4{,}65 \times 10^{-6}$
	$4{,}10 \times 10^{-6}$
3	$3{,}71 \times 10^{-6}$
	$3{,}67 \times 10^{-6}$
	$3{,}66 \times 10^{-6}$
	$3{,}99 \times 10^{-6}$
	$3{,}71 \times 10^{-6}$
arithmetic mean	$4.85 \pm 1.19 \times 10^{-6}$

Appendix D: Determination of the k_f-value (coefficient of hydraulic conductivity) of the sediments of Lake Müggelsee after DIN 19683-9; 1998 (Fachgruppe Wasserchemie in der Gesellschaft Deutscher Chemiker in Gemeinschaft mit dem Normenausschuß Wasserwesen (NAW) im Deutschen Institut für Normung e.V., 1999).

Acknowledgements

This thesis has only been possible with the help of many colleagues whose support and contributions I wish to acknowledge here.

Most of all I would like to thank my supervisors J. Lewandowski, M. Hupfer and G. Nützmann, who were the initiators of this project. Without their support, suggestions and professional advices, this work never would have been possible. I like to thank them especially for all their practical help, for constructive discussions and for their patience and assistance with reference to publishing our papers.

Furthermore, I would like to thank all colleagues and institutions who have been cooperating in the various subprojects of this work. The outcome of these subprojects has contributed significantly to this cumulative thesis. Therefore, I like to thank all of my co-authors, R. Buchert, F. Wilke and W. Brenner (UKE), R. Bergmann (FZD), A. Khalili and M. R. Morad (MPI). Especially, I would like to thank N. Hette, since his diploma thesis was a relevant part of this thesis and had a great influence on chapter II.V. Besides, I like to thank H.-J. Malitte, and his colleagues (BAM) for X-ray analysis.

This thesis would not possible without the help of many colleagues from the IGB. Many thanks go to S. Jordan for all her help and patience with sampling and experiments, E. Hamann for discussions and modelling my data, C. Dziallas for the analysis of the microbiological community structures presented in chapter III.III, C. Sturm for her help with several biogeochemical analysis, as well as C. Herzog, and all other members from the chemical laboratory of the IGB who supported me with analysis and in biogeochemical questions. Furthermore, I like to thank X.-F. Garcia for the larvae identification, K. Pohlmann for her help in statistical questions, B. Schütze and R. Hölzel for technical support with sampling and experimental setup, as well as S. Burkert for checking my English. Also I would like to thank the administrative stuff, all students and other colleagues who helped me somehow or other.

Last of all, I like to say thank you to all of those in my private environment who contributed, assisted, and diverted me during the last four years.

Supplement

Publications & presentations

Publications & presentations

Paper

Roskosch, A.; Hette, N.; Hupfer, M.; Lewandowski, J. (2012): Alteration of Chironomus plumosus ventilation activity and bioirrigation-mediated benthic fluxesby changes in temperature, oxygen concentration, and seasonal variations. *Freshwater Science*: in press

Roskosch, A.; Hupfer, M; Nützmann, G. & Lewandowski, J. (2011): Measurement techniques for quantification of pumping activity of invertebrates in small burrows. *Fundamental and Applied Limnology*: 178(2): 89-110

Morad, M. R.; Khalili, A.; Roskosch, A. & Lewandowski, J. (2010): Quantification of pumping rate of Chironomus plumosus larvae in natural burrows. *Aquatic Ecology* 44(1): 143-153

Roskosch, A.; Morad, M.R.; Khalili, A. & Lewandowski, J. (2010): Bioirrigation by Chironomus plumosus: advective flow investigated by particle image velocimetry. *Journal of the North American Benthological Society* 29(3): 789-802

Roskosch, A.; Lewandowski, J.; Bergmann, R.; Wilke, F.; Brenner, W, & Buchert R. (2010): Identification of transport processes in bioirrigated muddy sediments by [18F]fluoride PET (Positron Emission Tomography). *Applied Radiation and Isotopes* 68: 1094-1097

Extended abstracts & other articles

Roskosch, A.; Dziallas, C.; Grossard, H.-P.; Hupfer, M. & Lewandowski, J. (2009): Die Auswirkung von Bioirrigation auf mikrobielle Gemeinschaften und Prozesse in limnischen Sedimenten. Werder, *Erweiterte Zusammenfassung der DGL-Jahrestagung 2009*

Roskosch, A. (2009): Kleine Larve große Wirkung. IGB Jahresforschungsbericht 2008

Roskosch, A.; Hamann, E.; Nützmann, G.; Hupfer, M. & Lewandowski, J. (2009): Ecohydrological effects of *Chironomus plumosus* larvae on lake sediments. In: *Proceedings of the 7th International Symposium on Ecohydraulics*. Concepcion, Chile

Roskosch, A.; Jordan, S.; Hette, N.; Buchert, R.; Khalili, A.; Morad, M. R.; Nützmann, G.; Hupfer, M. & Lewandowski, J. (2009): Die Wirkung von *Chironomus plumosus* (Diptera: Chironomidae) auf Transportprozesse in limnischen Sedimenten. Werder, *Erweiterte Zusammenfassung der DGL-Jahrestagung 2008*

Seibt, C.; Hamann, E.; Roskosch, A.; Nützmann, G. & Lewandowski, J. (2009): Modellierung der von Chironomiden induzierten Austauschprozesse zwischen Sediment und Freiwasser. Werder, *Erweiterte Zusammenfassung der DGL-Jahrestagung 2008*

Roskosch, A.; Hupfer, M.; Nützmann, G. & Lewandowski, J. (2008): Messung von Fließgeschwindigkeiten und Pumpraten in Wohnröhren von *Chironomus plumosus*. Werder, *Erweiterte Zusammenfassung der DGL-Jahrestagung 2007*, 39-43.

Roskosch, A.; Hupfer, M.; Nützmann, G. & Lewandowski, L. (2007): Flow velocities and rates in burrows of *Chironomus plumosus* (Diptera: Chironomidae) in lake sediments. In: *Proceedings of the 11th Workshop on Physical Processes in Natural Waters*, Berlin, Berichte des IGB, Heft 25, 179-186.

Lewandowski, J.; Roskosch, A.; Hupfer, M. & Nützmann, G. (2007): Hydraulic and biogeochemical impacts of chironomids on nutrients. In: *Proceedings of the 6th International Symposium on Ecohydraulics*. Christchurch, New Zealand

Oral presentations

Roskosch, A.; Jordan, S., Hupfer, M.; Nützmann, G. & Lewandowski, J. (2010): Bioirrigation induced fluxes in sediments and their effects on nutrient distribution. Santa Fe, USA, ASLO Summer Meeting, 06.-11.06.2010

Roskosch, A. (2010): The effects of chironomids in sediments. Köln, Universität, Biowissenschaftliches Zentrum, Zoologisches Institut, Allgemeine Ökologie 14.04.2010, Posterpreisvortrag

Roskosch, A. (2009): Kleine Larve – große Wirkung! Wie Mückenlarven ein Gewässer beeinflussen. 2. IGB Wissenschaftstag, 19. 06.2009

Roskosch, A.; Hamann, E.; Nützmann, G.; Hupfer, M. & Lewandowski, J. (2009): Ecohydrological effects of *Chironomus plumosus* larvae on lake sediments. Concepción, Chile, 7th International Symposium on Ecohydraulics, 12. – 15.01.09, Vortrag.

Roskosch, A.; Khalili, A.; Hamann, E.; Buchert, R. & Lewandowski, J. (2009): Impacts of macrozoobenthos on hydrodynamic processes at sediment-water interfaces. Nice, ASLO Summer Meeting, 26. – 30.01.09

Roskosch, A.; Lewandowski, J.; Hupfer, M. & Nützmann, G. (2008): Impacts of *Chironomus plumosus* larvae on processes in lake sediments. St. John's, Newfoundland & Labrador, ASLO Summer Meeting, 08. – 13.06.08

Roskosch, A.; Jordan, S.; Hette, N.; Buchert, R.; Khalili, A.; Morad, M. R.; Nützmann, G.; Hupfer, M. & Lewandowski, J. (2008): Die Wirkung von *Chironomus plumosus* (Diptera: Chironomidae) auf Transportprozesse in limnischen Sedimenten. Konstanz, DGL Jahrestagung, 22. – 26.09.08

Roskosch, A. (2008): Der Einfluss von *Chironomus plumosus* Larven auf Transport- und Austauschprozesse in limnischen Sedimenten. Bad Saarow, BTU Cottbus, 13.11.08, Posterpreisvortrag.

Roskosch, A.; Hupfer, M.; Nützmann G. & Lewandowski, J. (2007): Flow velocities and rates in burrows of *Chironomus plumosus* (Diptera: Chironomidae) in lake sediments. Warnemünde, 11th Workshop on Physical Processes in Natural Waters (PPNW), 03. – 06.09.07

Lewandowski, J.; Roskosch, A.; Hupfer, M. & Nützmann G. (2007): Hydraulic and biogeochemical impacts of chironomids on nutrients. Christchurch, New Zealand, 6th International Symposium on Ecohydraulics, 19.02.07

Poster presentations

Roskosch, A.; Dziallas, C.; Grossar, H.-P.; Hupfer, M. & Lewandowski, J. (2009): Die Auswirkung von Bioirrigation auf mikrobielle Gemeinschaften und Prozesse in limnischen Sedimenten. Oldenburg, DGL Jahrestagung, 28.09 – 02.10.09

Seibt, C.; Hamann, E.; Roskosch, A.; Nützmann, G. & Lewandowski, J. (2008): Modellierung der von Chironomiden induzierten Austauschprozesse zwischen Sediment und Freiwasser. Konstanz, DGL Jahrestagung, 22. – 26.09.08

Buchert, R.; Roskosch, A.; Wilke, F.;Apostolova, I.; Brenner, W.; Lewandowski, J. & Khalili, A. (2008): Untersuchungen zum Einfluss von *Chironomus plumosus* auf die Hydrodynamik in Seesedimenten mittels PET/CT. Leipzig, 46. Jahrestagung der deutschen Gesellschaft für Nuklearmedizin, 23. – 26.04.08

Roskosch, A.; Hupfer, M.; Nützmann, G. & Lewandowski, J. (2007): Messung von Fließgeschwindigkeiten und Pumpraten in Wohnröhren von *Chironomus plumosus*. Münster, DGL Jahrestagung, 24. – 28.09.07

i want morebooks!

Buy your books fast and straightforward online - at one of world's fastest growing online book stores! Environmentally sound due to Print-on-Demand technologies.

Buy your books online at
www.get-morebooks.com

Kaufen Sie Ihre Bücher schnell und unkompliziert online – auf einer der am schnellsten wachsenden Buchhandelsplattformen weltweit! Dank Print-On-Demand umwelt- und ressourcenschonend produziert.

Bücher schneller online kaufen
www.morebooks.de

VDM Verlagsservicegesellschaft mbH
Heinrich-Böcking-Str. 6-8
D - 66121 Saarbrücken

Telefon: +49 681 3720 174
Telefax: +49 681 3720 1749

info@vdm-vsg.de
www.vdm-vsg.de

Printed by Books on Demand GmbH, Norderstedt / Germany